Uncertainty Bands: A Guide to Predicting and Regulating Economic Processes

T0247814

Uncertainty Bands:
A Guide to Predicting and Regulating Economic Processes

Ashot Tavadyan

ANTHEM PRESS

Anthem Press

An imprint of Wimbledon Publishing Company
www.anthempress.com

This edition first published in UK and USA 2022
by ANTHEM PRESS
75–76 Blackfriars Road, London SE1 8HA, UK
or PO Box 9779, London SW19 7ZG, UK
and
244 Madison Ave #116, New York, NY 10016, USA

British Library Cataloguing-in-Publication Data
A catalogue record for this book is available from the British Library.

Library of Congress Control Number: 2022932213

ISBN-13: 978-1-83998-398-6 (Pbk)
ISBN-10: 1-83998-398-1 (Pbk)

Cover credit: Photograph by Aghasi Tavadyan

This title is also available as an e-book.

CONTENTS

Preface   vii

Introduction: The Philosophy of Economic Forecasting   1

1   Interval Links in Economy and the Capabilities of
     Quantitative Thinking   7
     §1. The Interval Method and Systemology   7
     §2. The Quantitative Thinking and the Economic Arrhythmia   11
     Summary of Chapter 1   14

2   The Possibilities for Forecasting Economic Indicators   17
     §1. The Effect of Compressed Spring   17
     §2. The Uncertainty Relations in Economy   21
     Summary of Chapter 2   24

3   The Principle of the Minimal Uncertainty Interval   27
     §1. The Minimal Uncertainty Intervals of Economic Indicators   27
     §2. The Effect of the Expanding Uncertainty Bands in Economy   33
     §3. The Sensitivity Thresholds of Economy   35
     Summary of Chapter 3   38

4   The Intervals of Key Economic Indicators   41
     §1. The Systematization of Economic Indicators   41
     §2. Intervals for Target and Regulatory Indicators   44
     §3. The Key Macroeconomic Task   50
     Summary of Chapter 4   60

5   Key Principles of Economic Regulation   63
     §1. Economic Diseases Caused by Regulation   63
     §2. Ten Principles for Forecasting Economic Processes   70
     Summary of Chapter 5   76

Conclusion   79

Appendix: The Uncertainty Relations of Economic Indicators   83

Acknowledgments   87

Index   89

PREFACE

The role of uncertainty has increased in the economy. *The economy shows increased volatility; the frequency of changing situations and policies can rapidly change the economic landscape, thus giving them a new quality.*

An economy does not develop along with programmed principles. The economic, social and political situations and the market, as well as international economic relationships, are variable. Economic regulation oftentimes requires substantial adjustments demanding readiness, which can be achieved using the intervals of economic indicators.

In a transforming economy, the issue of uncertainty is even more urgent, while the aftermath of unpredictability and an unfeasibility of precision forecasting the economic processes may be even more sensitive. The transforming economies are the states staging the process of transforming the command economy into the market economy. The economic regulation constantly changes in those countries. This is reflected in the key economic indicators characterizing the country's development. The level of uncertainty is higher, especially in transforming economies, and the likelihood of forecast fulfillment is lower. Uncertainty intervals for forecasting are wider in those countries as they have more sensitivity thresholds, which are the critical bounds of key economic indicators. The crossing of those thresholds can cause even greater changes than in the developed countries, which have stronger homeostasis of the economic system. For instance, the drop of GDP in transforming economies after the crisis by far exceeds the drop of GDP in developed countries.

Under those conditions, it is rather unproductive to compile static equations to be used in studying the economic processes. Moreover, the number of sensitivity thresholds increases, which changes the elasticity of economic indicators. Besides, the same value of an indicator, say, inflation or an exchange rate, may correspond to very different values of other economic indicators such as GDP, its structure and exports. Thus, those causal links are ambiguous; hence, the forecasts compiled upon unambiguous links will be inadequate for describing economic prospects.

The principles of the economy are mostly identical; however, they are manifested with differing intensities under different conditions and times. Economic diseases, having a universal character, show themselves distinctly in transforming economies. It is especially a small ship coming to be in the process of renovation in heavy seas that encounters quite some unexpected situations. Modern ships may also encounter those problems in very heavy seas. The seas of uncertainty have become more turbulent for all, and the detection of relevant problems related to uncertainty and their research methodology has gained priority for all countries.

The minimal uncertainty intervals thus presented are larger in the countries with a higher level of instability. Therefore, wider intervals will oftentimes include substantial negative processes. That is why we must continually analyze the impact of uncertainty on key economic processes.

The need to write this book had arisen in the aftermath of studying the factors of uncertainty that seem to possess a universal economic character. One of the leading stimuli for writing this book has been an effort to identify the uncertainty intervals that are present in all economic processes and often are particularly pronounced in transforming economies.

This book presents the system of interconnected key indicators. The matter is that when formulating each connection there may occur particular lapses, which as a whole significantly distance the forecast from the reality. The economic environment is subject to abrupt changes ever-increasing in numbers and impact, thus multiplying the probability of forecasting error. Meanwhile, uncertainty intervals play a positive role in predictions by raising their probability.

A precise forecast may considerably amplify the aftermath of critical processes. The pursuit of precision may result in very negative results. In this case, less attention is devoted to the economic cushion, for a precise forecast displays the economy as if all details are exactly determined. A precise forecast converts the forecast into a lottery with all its negative features, whereas the most likely results could be dumped.

An array of conclusions has been herein formulated upon economic indicators being within certain uncertainty intervals, which dynamically become uncertainty bands. This book shows their real validity for economic research.

This book is leaning upon constructive thinking, including quantitative. For scientific modeling, a way of thinking is in the first place, rather than just a device for formalizing the assignment. This is a method for identifying the problems and the process of their exploration. An exposition of the problem, its diagnosing, is a key step for exploring causations of the economy.

This book will help a reader to generate a system-oriented image of uncertainty in the economy and the capabilities of economic forecasting. It can develop approaches to analysis and forecasting the economic processes and may become a basis of breakthrough decisions in the economy.

INTRODUCTION: THE PHILOSOPHY OF ECONOMIC FORECASTING

The main objective of this work is to study the intervals of key economic indicators facilitating economic growth under conditions of uncertain economic processes. The methods herein applied for both the evaluation of key causalities of indicators and the principles of economic forecasts have been stipulated by the factor of uncertainty and by the need for systemic analysis of economic interdependencies. It should be noted that the complexity of the model, as will be demonstrated in this book, has no decisive role in uncovering the main principles of key economic causalities. Based upon the synthesis of economic research and the analysis of statistical data, this book presents a relevant and quite illustrative approach for indicator systematization and interval research. The economy, as any low-validity system, is inherently volatile and hard to predict, making it difficult to unambiguously point out the causalities of an economic system. In this context, determining the interval of indicators is most productive within the bounds in which their values have the highest likelihood.

The book cites a complex analysis of key economic indicators under conditions of uncertainty, which is the natural state of the economy. It clarifies the essence of economic indicators and substantiates the need for certain devices based on their systemic analysis. Any proposal should not logically contradict the causalities of the indicator system. The systematization of economic indicators is a synthesis between the analysis of economic causalities of indicators and the solutions of specific issues in the economy.

The real situation requires a study to be made not only of the state of equilibrium under growth but also mainly of its malfunction, with due regards to the fact that in the economy the unforeseen events may have substantial, even negative aftermaths. Only complex research of the system of economic indicators under uncertainty will help to formulate the principles to be achieved, which will yield adequate solutions.

There exists a widespread, somewhat vulgar notion on the connection between economic research and economic practice that any economic research has to deliver an inventory of specific instructive practical recommendations, and vice versa, any efficient action in economic regulation has to be substantiated with scientific conclusions. This type of straightforward connection is erroneous. It is not there even in the renowned theory of Keynes. Therefore, this book highlights the main directions of impact on the economic practice, shows its role in identifying substantial causal links of economic indicators and exposes the key economic connections and the main instruments of affecting them.

The reality presents complex requirements for determining the minimal intervals of the uncertainty of economic indicators. Nonetheless, the presented intervals, uncertainty bands and the uncertainty ratio, as well as sensitivity thresholds, can become a basis for studying the real causalities of economic indicators, considering all possible adjustments introduced by the actual world to the indicated methods of studying the economic indicators.

The purpose of the work is in elaborating the interval methods of studying the key economic indicators under uncertainty as well as in their productive systematization enabling to visualize the prediction capabilities and the variability of economic processes.

To achieve the set goal, this book explores the issues presented below.

***

This book explores the philosophy of economic modeling under uncertainty. Fatal flaws are possible in economic researches. To overcome possible flaws, there is a poor practice to quantitatively depict the economic relations by implementing models with more and more growing complexity. However, one cannot easily build tall structures on the quicksand of ever-changing economic data. In analogy to the simplification principle of Occam's razor, it is expedient to formulate the following principle of growing complexity: if increasing the complexity of a model does not improve the validity of its assumptions, while its results are not verifiable in accordance to the reality, then the growing complexity can hardly be viable.

In economy the extreme events are difficult and often impossible to predict, especially when leaning upon the past alone. Events can be approximately evaluated, as they may be sudden and erratic.

It is imprudent to plan and construct colossal and rigid structures where the economic landscape constantly alters, and the foundation may shift. Models and forecasts, as buildings, must be built in consonance with the landscape. It is misleading to take a part of the structure that works in a tropical setting, then blindly copy and implement it in subarctic regions and vice versa. In this sense, attempting to study an indicator outside its system, with no actual regard for the systemology, will result in a misrepresentation of the causal links in the economy, thereby exaggerating a specific economic connection, thus distorting even the generalized economic assumptions.

When constructing the model, the pieces that make up the foundation must not be ignored. The key indicators are the cornerstones of the economy, for economic activity cannot be efficiently implemented if the system of indicator interdependencies is not identified, those indicators being characteristic of the determinant economic connections.

***

Many reports that are aimed to alter economic policies use some kind of a model to justify their means. In this case, models serve as some kind of confirmation of the quality of said reports. Usually, reports or articles implementing a model have a bigger impact than those that do not have one. If a model only serves as a requirement to justify a report or a strategy, then it is self-deception that can outgrow into policy delusion. In those cases, data is oftentimes simply gathered by the principle of accessibility. Accessible, known data is used to justify the means.

A model should not be an end in itself. Its use must be justified and compared against alternatives, and the results should be validated. Besides, a precise forecast, especially in crises, significantly reduces its fulfillment possibility and does not contribute to the creation of a mechanism for adjusting an economic policy.

The inflexible, precise forecast gives the illusion of stability. A seemingly stable situation is by no means equivalent to a low-risk situation of negative consequences, especially in unforeseen circumstances. Moreover, the calculated mean or the average often distorts reality. The concept of the average is inefficient; the key here is the minimal uncertainty interval, which will be discussed in this book. A scenario analysis should be present in any interval forecast. This approach is much more efficient than a precise forecast.

Business cycles are in an uncertainty band, and the projections of business cycles are becoming progressively provisional because economic crises are increasingly unpredictable both in time and in depth. After some time, the forecast quality changes, and it becomes artificial. The economic indicator varies in an uncertainty band over time, wherein various scenarios are possible. Quantity always can change to a new quality.

It is a norm to define several variables as explanatory to construct some models. However, the initial isolation of several variables from the system is unacceptable because it immediately breaks economic causalities. Obtained relations between economic indicators are insufficient. Those results will be optimal for the model, but they may not be optimal for the real system of economic relations.

The economy is a living system that must ensure smoothing out under stress. When the economic policy becomes stiff because of rigid models, the economy itself becomes rigid. In this case, with the accumulation of negative phenomena and with the transition of their quantity to a new quality, the spring of accumulated problems may be sharply released.

***

Phase transition can be described as a process of transition between the basic states of a system. During a phase transition, certain properties of the system change, often drastically. This book reasons that the economic system has sensitivity thresholds—critical states of economic processes at which they drastically change their characteristics. In those critical states, economic phase transitions may occur. For example, a slowdown in economic growth, especially a tangible recession, can result in negative qualitative changes in the economy, which can have a long-term impact.

In this sense, the identification of the sensitivity thresholds of the economy contributes to the assessment of economic homeostasis. Depending on the range of uncertainty intervals, an appropriate toolkit should be formulated against untoward events to prevent a negative impact on economic processes. When developing this toolkit, it should be borne in mind that uncertainty tends to expand over time. The presentation of programs for years ahead is of little value because the uncertainty intervals expand when economic volatility increases, forming uncertainty bands. You can find the effect of an expanding uncertainty band in this book.

To solve the issue of inflexible programs, this book presents the minimal uncertainty interval. This method aims to find a logical path from the predicament of explaining its quantitatively precise indefinability of the indicators and helps to deploy a complete picture of causal links in the economy. The aim is to contribute to the flexible and realistic concept about possible dynamics of economic processes with interval forecasts and probabilistic evaluations of those events' outcomes.

***

Models may provide hope that economic indicators are precisely defined and constant. This is far from reality and wishful thinking. Each of us when given the same pallet of indicators and toolkit of models to give an analysis on some economic process may paint different scenarios. Economic assessments and forecasts contain an element of art. This involves finding effective solutions in acceptable ranges of economic indicators.

The art of economic regulation lies in the determination of acceptable intervals for regulatory indicators to overcome the positive sensitivity thresholds of target indicators, given the normative indicators remain in their favorable intervals. To do so, key indicators are systematized, and their uncertainty intervals are identified. This systemic approach should be carried out in combination with economic objectives, normative constraints and the basic principles of regulation in the economy.

To assist in painting a desirable economic scenery, this book formulates the key macroeconomic task based on the intervals of key indicators. It presents the fundamental goal of economic development, restrictions on normative indicators and conditions on regulatory indicators. The key macroeconomic task makes it possible to overcome the complexity of the systemic representation of the terms ensuring sustainable economic development.

***

In both medicine and economics, the dose of a medication or the value of an indicator can vary from case to case and should be considered within an interval. The subscribed dose depends on the patient's condition, the course of the disease and the qualification of the doctor. So, in economics, the choice of the optimal interval for an indicator depends on the state of the economy, the dynamics of the economy, and, of course, the qualification of a decision-maker. In both sciences, there is always an element of unexpected, thus a checklist of principles should be put in place.

It is indicated that exiting the allowable intervals of regulatory indicators contributes to the emergence of economic diseases. This book tries to explore and systematize economic diseases, presents the factors that affect forecast efficiency and makes the forecast satisfactory.

As in medicine, in economics, the diagnosis must not be rigid and permanent. Vital indicators could and will change throughout therapy. A rigid forecast resulting in an attempt to ensure the rigid stability of indicators is dangerous and fraught with irreparable economic losses. With a point forecast, the economy is adjusted to the chosen model, and not vice versa. This in effect turns the model into a Procrustean bed for the economy, which often results in a significant drop in GDP and degradation of the economic structure. As with vital indicators, each key economic indicator should vary in its optimal range.

Based on the systematization of the conducted research, the final chapter formulates the ten principles of forecasting, which are necessary for forecasting the economic processes and decision making under uncertainty.

Systemic arrangement of causations of economic indicators under uncertainty, natural for the economy, enables to advance many new important statements on the methods of evaluating the economic processes and the role of quantitative analysis in the economy. Of direct practical significance are both the presented intervals of a target, normative and regulating indicators, and the formulated key macroeconomic task, as well as the principles of forecasting.

Naturally used in the research was economic literature presenting detailed causalities of indicators. As a matter, the given analysis of indicator connections has been compared with the reality of both positive and negative experiences of making economic decisions. Moreover, there is no need to particularly derive the causalities of indicators being quite demonstrative and adopted by all representatives of economic theory and practice. In this regard, the study indicates the given causal links with no special analysis. For economic indicators using differing names, use was made of the most applicable ones. It should be noted that the inter-indicator dependences are diversified, when indirect and direct causes were present, preference was given to their direct links.

Chapter 1

INTERVAL LINKS IN ECONOMY AND THE CAPABILITIES OF QUANTITATIVE THINKING

§1. The Interval Method and Systemology

Protection from the misrepresentation of economic indicators

A real process may not be subject to measuring, nor be exactly measured, for the economic situation is continually subject to critical alteration, with periodical considerable perturbations.

It is to be recognized that there are no methods for precise measuring of the economic processes. Thus, when quantitatively formulating the precisely indeterminate processes, it must be identified whether its evaluation is distorted. The interval-based method is particularly useful for processes that cannot be precisely measured or predicted due to subjective or even objective reasons, for it will enable to describe the process while discovering its essence with no distortions.

A point estimate of an economic indicator often results in the loss of several of its attributes; hence, the book presents the interval method enabling quantifying the process with minimal losses in determining the indicator attributes. It is the indicative estimation of the process attributes manifested in the reality that will give a systemic picture of the quantitative and qualitative features of said process.

The interval method will enable a most complete description to be made of the essence and the manifestation of the real processes; hence, fully satisfying is the principle of Occam's razor which can be regarded as a sufficient condition for the application of this method.[1] Given the impossibility of accurate

---

1 The most known depiction of Occam's razor principle is attributed to Albert Einstein. It says: "Alles soll so einfach wie möglich sein, aber nicht einfacher" (Everything should be made as simple as possible, but not simpler). *The Ultimate Quotable Einstein* (Princeton: Princeton University Press, 2010).

measurement in the economy, the interval method allows us to overcome the complexity of describing economic processes.

The interval determination of indicators is a method of research for precisely indeterminate indicators under the state of uncertainty natural for the economy. The interval representation of the indicator will enable a description to be made of its most likely values. The interval method of economic indicators will open a logical path from the predicament of explaining its quantitatively precise indefinability. It will help to deploy a complete picture of causal links in the economy.

The capabilities of the existing methods of evaluation and forecasting the economic indicators cannot be seen as an absolute cure-all. This is no less hazardous than their complete disregard. Meanwhile, for an efficient quantitative analysis of economic processes, it is necessary to determine the content of economic causal links and the dependencies of the indicators. Only after that is done the quantitative analysis will become coherent to the essence of reality and yield meaningful results.

Quantitative changes may occur even if the essential attributes of reality are identical. This once again emphasizes the expedient nature of the interval method. The method also specifies the quantitative features based on the essential attributes underlying any study. This makes the interval method an efficient tool when analyzing the economy.

The interval method of analyzing the economic processes enables both the optimal combination of qualitative and quantitative methods of exploring the economic indicators and their systematization.

The systematization of economic indicators as one of the fundamental problems of the economy has as of today been studied insufficiently. The objective reason is the complexity of the subject. The main subjective reason is an incomplete agreement between the quality of research and mathematical modeling.

An important task to be resolved along with bringing the indicators into a single system is the analytical identification of the framework shaping the system of economic indicators. This is the process of systemology.

Exploring the system of key economic indicators is a priority practical requirement because perfecting the economic regulation is unlikely without considering the substantial interdependencies of economic processes under the dynamic interaction. Analyzing the empirical connections and identification of substantial economic links will enable the system of key economic indicators to be presented. This exploration of economic causations identifies their substantial links and estimates possible values. The determination of the main causal links of the system forms the basis of the systemic presentation of economic indicators.

The fundamental principle of exploring an economic system is to determine its key indicators, to identify and systematically arrange the intervals of their possible values, using the art of regulation.

Identifying the substantial causations does not mean ignoring the practical results. The practical results are tangible and enable the clarification of the key causal links as well as the verification of their correct formulations. The substantial causations and practical results are in a systemic unity containing an integral representation of reality. Thus, economic research should also include scientifically digested substantial causations herein presented as intervals of key indicators. Those intervals will be systematized.

It must be admitted that the true essence of the economy remains hidden if a study does not explain the real economic phenomenon. In this case, the knowledge reflecting the essence of the economy represents no objective truth. For example, multiple unsuccessful attempts have been made to calculate the exact price of a commodity. Alas, an exact price cannot be determined. That result, significant in itself, will enable argumentative deductions to be also derived about other economic processes.

From here the question of the compliance of the presented key connections to their true causations follows. Albeit those connections are verified for validity through practice, where the accuracy of theoretical deductions is confirmed by the sufficiently correct forecasting of the observed results. It will be shown that it is reasonable to predict the substantial or key processes only within an interval.

A causal link should not exclude chance and uncertainty in the economy. Therefore, formulating the targets only through the identified connections may subsequently produce a discrepancy between the genuine and the represented economic processes. Hence, when formulating the objectives, one must proceed not only from the causal links but also from the uncertainty within the economic system.

The interval representation of the economy plays an important role in considering their causal links because real processes do not reflect the substantial economic links. Given the factor of uncertainty, the interval method contributes to the complete presentation of economic interdependencies without substantial losses of the most probable values of economic indicators.

The systemic approach to analyzing the economic indicators, first of all, entails a study of the key indicators as the interconnected structural parts of a unified system of economic indicators; secondly, the systemic approach means the determination of the role of each indicator within the economic process. The system of key indicators will enable a transition to a sufficiently complete understanding of economic processes.

Systemology is the determination of connections between the parts of a system, their description and the classification of the identified economic patterns. A systematization of indicators is necessary for all divisions of academic knowledge exploring complex subordinated processes. An implementation of the systemic approach enables a demonstrative description to be made of an economic indicator's substantial causations, their structural positions concerning other indicators, as well as clarifying the key connections and assessing the system of causal links. It will also contribute to the identification of unknown systemic connections of economic processes. The systemology of economic indicators is essential for economic practice, since it identifies the basic economic connections, making the indicator connection analysis accessible and convenient.

Attempting to study an indicator outside its system, with no actual regard for the systemology, will result in a misrepresentation of the causal links in the economy, thereby exaggerating a specific connection, thus distorting even generalized economic assumptions.

The system of key economic indicators represents the integrity of interconnected indicators standing for the substantial facet of the economic causations system, the part of which cannot be predetermined using the other, regardless of being interconnected in the single economic system.

The systemology of economic indicators makes it possible to establish their unity, to identify their causalities, to understand at a deeper level the essence of economic processes. A system is larger than a mere sum of its parts. In this context, causality and uncertainty are two sides of the same coin and must be regarded as a system, especially when exploring the interconnections of the key economic indicators. Systemology demands a description of the substantial features of the economic system, its structure, ramified chain reactions, the weak links and the factor of uncertainty. An economic system will more often than not change places of cause and effect, especially during critical situations.

Elaborating on the methodological problems of this approach has resulted from the generally recognized significance of systemic analysis. Those problems count a substantial number of publications. Therefore, rather than targeting a wide exploration of systemic presentation, the book is an effort to discover the logic of interconnections of economic indicators.

Understanding the importance of systems science has brought about efforts to formalize this concept in different fields of knowledge. It is to be noted that this formalization is effective as soon as it displays a substantial aspect of the knowledge of reality.

When analyzing an economic system, it is necessary to identify the elements determinant of a given system's essential attributes, since the identification of those attributes within an economic system will yield the definition of the structure of the given system, as well as the specific causations of its components. The efficiency of systemic thinking is in the first place an ability to identify the key connections within a system, which are the generators of the system's structure and causations.

The study of key indicators is needed for analyzing the causations and for depicting a certain part of an economic system, which it characterizes in a generalized way. The key indicators are the hub for the remaining variables in an economic system, and their essential properties are uncovered through analyzing those leading indicators.

The key system-generating indicators are the cornerstones of the economy, for economic activity cannot be efficiently implemented if the system of indicator interdependencies is not identified, those indicators being characteristic of the determinant economic connections.

§2. The Quantitative Thinking and the Economic Arrhythmia

The Procrustean bed for quantitative methods

Given the issue, economic studies make extensive use of models having differing degrees of aggregated economic information using differing structuring methods and differing levels of approximating the economic reality. Meanwhile, the results obtained can promote the solutions to economic problems, provided the study is accompanied by a substantive analysis of both the assumptions and the obtained result.

Using any type of model will anyway invoke a host of assumptions. However, an analysis may be justified in case it exposes substantial economic causations. The problem arises when selecting the types of models for describing the real causations and in suggesting assumptions.

The more demonstrative the model showing the real economic causations is, the easier it is used for both theoretical analysis and specific calculations. Though, the simplification of the model assumptions may cause a divergence from the current economic process. In this case, certain limitations of the research results should be considered. On the other hand, the more complete the model's presentation of the complex economic causations is, the more difficult it becomes to use it for real-world calculations. Paradoxically, the value of those models decreases.

A model elaboration that is diverted from reality can more often than not go further into "art for art's sake." The task has to be delineated before developing the models to describe it.

There is hardly any sense for it to be done in reverse—to squeeze the reality into the Procrustean bed of however complex quantitative models. In turn, model complexity can oftentimes cover its inconsistency with the real world. As to the analysis of economic reality yielding reasonable results, it is commonly based upon relatively simple models, which give meaningful results with no substantial interior simplifications causing noticeable diversions from reality.

Complex methods will not necessarily produce a substantial reduction of uncertainty intervals. S. Makridakis and M. Hibon emphasize that the newest and most complex methods are not bound to ensure more accurate results when compared to the simple ones.[2] When using models, the attention is mostly focused on functional cases but seldom on inconsistent examples.[3] As a rule, the accent is made on steady-state results disregarding the factor of uncertainty in multiple cases of discrepancies in dynamic situations that rarely fail to accompany the economic processes. One can never disregard the research results showing that the simpler forecasting and solution-searching methods work better than the complex ones.[4] S. Armstrong also finds that the more complex the model, the worse the forecast.[5]

That is quite substantiated, for, in that case, it is more difficult to give a distinct description of patterns and causal links in the economy as well as to amass the needed information. Besides, uncertainty is an objective feature of an economic system, as will also be shown by the uncertainty relations. A growing complication of a model does not always improve its quality, although it may impede the capacity to evaluate the economic processes. An analogy with medicine would be appropriate here. It is better to use drugs with known effects than those with unknown effects.

One should not only consider the extent of the possible use of the model, though it matters as well. In analogy with the simplification principle of Occam's razor, it is expedient to formulate the following principle of growing complexity: if increasing the complexity of a model does not improve the

2 S. Makridakis and M. Hibon, "The M3-Complication: Results, Conclusions and Implications," *International Journal of Forecasting* 16 (2000): 451–76.

3 Nassim Nicholas Taleb, *The Black Swan: The Impact of the Highly Improbable* (New York: Random House, 2007).

4 Robyn M. Dawes, *Everyday Irrationality: How Pseudo-Scientists, Lunatics, and the Rest of Us Systematically Fail to Think Rationally* (Westview: Routledge, 2001); Goldstein, D. G., & Gigerenzer, G. (1999). *The recognition heuristic: How ignorance makes us smart.* In G. Gigerenzer, P. M. Todd, & The ABC Research Group, Simple heuristics that make us smart (pp. 37–58). Oxford University Press.

5 Armstrong J. Scott, *Principles of Forecasting: A Handbook for Researchers and Practitioners*, J. Scott Armstrong (Ed.), (2001), Boston: Kluwer Academic Publishers.

validity of its assumptions, while its results are not verifiable in accordance to the reality, then the growing complexity can hardly be viable.

If an increased complexity of interconnections adds nothing to the main body of economic research, then building up the model complexity is unfounded: hence, the criterion of selecting the research method of uncertainty intervals. An increasingly complicated research method producing no reduction of the minimal uncertainty interval is hardly useful.

A quantitative economic model based on systemic analysis of causal links in the economy and on a scenario approach may facilitate a reduction of the uncertainty intervals; however, it cannot give precise solutions because of the uncertainty factor.

The reality is undoubtedly the best model representation of itself. All models are limited by the definition; they are not completely reflective of reality. An effort of building up the complication of models does not resolve this issue, while the research results will show substantial growth in complexity. Model accuracy from including additional factors may decrease at a certain point and add unnecessary complications. Besides, the validity of selected factors in quantitative economic models is oftentimes not substantiated, nor is there an explanation of their connections with the unselected factors. The evaluation of results always goes with a reservation: other things being equal, that is, other factors being unchanged. However, a change in one factor will as a rule change the causal links.

An economy is a complex system with obscured causal links. In this case, the extreme events are too difficult and often impossible to predict, especially when leaning upon the past only, wherein the critical events do not always end with crises. Events can be approximately evaluated, but they may often be sudden and erratic. The interval forecasts with probabilistic evaluations of those events' outcomes contribute to the flexible and realistic concept about possible dynamics of economic processes.

Research methodology including the quantitative economic methods is primarily a mindset, which contains the interval method. Problem formulation and comprehensive presentation of research methods are of greater significance than merely a model selection and its subsequent results. The primary category is a quantitative economic mindset, rather than a model.

The quantitative economic mindset makes it possible to formulate the objective and approaches to its optimization, to estimate the economic causations, to identify the bottlenecks of the economy, to analyze its risks and limitations, to approximate the assessment of uncertainty and to evaluate the interval forecasts and their probabilities. The economy is dynamic in its essential quality. Here precise forecasts are impractical and ineffective. Hence, the quantitative economic mindset makes it possible to evaluate the intervals of forecasts, to select its upper and lower bounds enclosing the forecasts that are most likely to be implemented.

Constructive thinking, including the quantitative economic mindset, makes it possible to delineate the principle of minimal uncertainty interval, the uncertainty bands and the sensitivity thresholds and to present the effect of the compressed spring, the uncertainty relations, the principles of systematization and forecasting the economic indicators, as well as the key macroeconomic task.

Summary of Chapter 1

§1. The interval method and systemology

- A real process may be not exactly measured in the economy, for the economic situation is continually subject to alteration, with periodical considerable perturbations. The interval method of economic indicators will open a logical path from the predicament of explaining the quantitatively precise indefinability of the indicators. It will help to deploy a complete picture of causal links in the economy.
- The fundamental principle of exploring an economic system is to determine its key indicators, to identify and systematically arrange the intervals of their possible values, using the art of regulation.
- The systemic approach to analyzing the economic indicators, first of all, entails a study of the key indicators as the interconnected structural parts of a unified system of economic indicators; secondly, the systemic approach means the determination of the role of each indicator within the economic process.
- Systemology is the determination of connections between the parts of a system, their description and the classification of the identified economic patterns. Attempting to study an indicator outside its system, with no actual regard for the systemology, will result in a misrepresentation of the causal links in the economy, thereby exaggerating a specific economic connection, thus distorting even the generalized economic assumptions.
- The system of key economic indicators represents the integrity of interconnected indicators standing for the substantial facet of the economic causations system, the part of which cannot be predetermined using the other, regardless of being interconnected in the single economic system.
- The study of key indicators is needed for analyzing the causations and for depicting a certain part of an economic system, which it characterizes in a generalized way. The key indicators are the cornerstones of the

economy, for economic activity cannot be efficiently implemented if the system of indicator interdependencies is not identified, those indicators being characteristic of the determinant economic connections.

§2. Quantitative thinking and the economic arrhythmia

- In analogy with the simplification principle of Occam's razor, it is expedient to formulate the following principle of growing complexity: if increasing the complexity of a model does not improve the validity of its assumptions, while its results are not verifiable in accordance to the reality, then the growing complexity can hardly be viable: hence, the criterion of selecting the research method of uncertainty intervals. An increasingly complicated research method producing no reduction of the minimal uncertainty interval is hardly useful.

- In the economy, extreme events are too difficult and often impossible to predict, especially when leaning upon the past only. Events can be approximately evaluated, but they may often be sudden and erratic. The interval forecasts with probabilistic evaluations of those events' outcomes contribute to the flexible and realistic concept about possible dynamics of economic processes.

- Research methodology including the quantitative economic methods is primarily a mindset. The quantitative economic mindset makes it possible to formulate the objective and approaches to its optimization, to estimate the economic causations, to identify the bottlenecks of the economy, to analyze its risks and limitations, to approximate the assessment of uncertainty and to evaluate the interval forecasts and their probabilities.

Chapter 2

THE POSSIBILITIES FOR FORECASTING ECONOMIC INDICATORS

§1. The Effect of Compressed Spring

The rules of the economic game are continually changing
or the unreliability of the past-based forecasts

Equilibrium can be dangerous. With the accumulation of negative phenomena and with the transition of their quantity to a new quality, the "spring" of accumulated problems may be suddenly released. An unexpected issue in the market causes the spring to rebound. Incorrectly implemented policies that mainly focus on one economic indicator wind the spring up. Paradoxically, small doses of deviations, in contrast to rigid stability, reduce the risk of reaching the threshold of critical fluctuation. Moreover, slight stress allows to identify the weak links in the system. It is highly advisable to denote that a seemingly stable economic situation is by no means equivalent to a low-risk situation of negative consequences, especially in unforeseen circumstances.

Econometric or other forecasting methods, based on data of even a long-term stable development or even a minor crisis, cannot, by definition, predict a future crisis caused by newly emerging factors, especially with high accuracy. In those cases, the examined variable is presented as endogenous, which means that the variable is determined by the model. The forecast is then carried out based on those values of selected economic variables, defined as exogenous, which are determined outside the model. The result is based on the analysis of observations of the past behavior and works only for the past, that is, for the already available data. However, the situation in the economy is volatile, sometimes subject to significant changes. The relations between variables based on already known data can always be formulated, although the forecast based on the past values of endogenous variables is rather uncertain.

In mathematical methods, the possibility of a significant economic change and the factor of uncertainty are often underestimated. At the same time, artificial patterns are revealed, and cause-and-effect relations are presented in an explicit form where they may not exist. Accepting an outcome after a certain procedure does not mean it has resulted from that same procedure, since the cause-and-effect relations are by no means chronological.

Econometrics can produce veritable results for the future only where the dynamic processes are not subject to significant changes. However, in economics, this assumption is utopian. Econometrics equalizes the outliers. As a result, all the important outliers that happen a priori may be unaccounted for. In general, if a model does not correspond to the real parameters of an economic indicator, then the relevant period of the indicator's earlier state is meaningless.

The most feasible substantive record of the patterns of an event is only a necessary condition, but due to the factor of uncertainty, it cannot be sufficient. Even with a certain correlation of the model to the processes under investigation, deviations are unavoidable. The problem is how to minimize deviations in assessing and predicting an event. To minimize this problem, it is suggested to do the assessment and generation of uncertainty intervals for target indicators and possible ranges of normative and regulatory indicators by using the sensitivity thresholds.

Given the chance, people are often inclined to take the pill when unwell without specifying the diagnosis. At the same time, they often imagine a disease that would fit an unconsciously chosen medicine. In economics, it is often the case that some forecasting models may be applied without proper testing or evaluation. However, models like drugs do not cure every disease. They cannot even help to improve the research process, but rather worsen the situation, especially with a flawed economic diagnosis. Can we understand what health is by turning a blind eye to a disease? Unfortunately, in economics, uneventful periods are often not important at all and do not play a role. If only uneventful periods are considered and the calculations are made in the times of change, then the probability of a particular event is attributed to a "scientific" definition. This creates an impression of being scientific where only scientism prevails.

This results in the fact that in uneventful periods when the role of the chosen model is less significant, the effect of extrapolation comes into play. In those periods the efficiency of the chosen model is justified, and there is no mention given that it is not suitable in traumatic economic situations. Significant crisis processes are unlikely, but their consequences can be very significant and painful. Moreover, they are difficult to evaluate.

Economic forecasts sometimes resemble old weather forecasts: tomorrow will be like today, with a negative deviation if the winter is coming, and with a positive deviation if the summer is nearby, without considering the key patterns changing the situation. Moreover, those forecasts are done without considering the factor of uncertainty and the possibility of a sharp change in the course of dynamic economic processes.

An illogical economic forecast moreover presented based on mathematics, in particular, econometrics, is not acceptable. However, a logical result, which at first glance seems to consider obvious cause-and-effect relations and possible feedbacks, still means little. The fact is that the logical result only with a caveat—"all other things being equal"—can be considered a necessary, but not sufficient, condition for the model. Quantitative analysis is initially limited, but to be useful, it must at least satisfy the condition of necessity—a meaningful assessment of the causal links in the economy.

The same value of one indicator, say, inflation, can correspond to completely different values of GDP, the balance of payments, the balance of trade, and employment, so the conformity here is ambiguous. Consequently, economic calculations are ambiguous and arbitrary. Whether the preliminary selection of several exogenous factors is correct is the first question, the second being how they relate to those which have not been selected.

The predictive capability of any model is overestimated. The practical impossibility of anticipating crises using models depends not only on the fact that the number of unknowns exceeds the number of equations formulated even in the most careful way but also on the fact that the result depends on the model itself, which is framed by a person or a group. This is of course subjective.

Stock exchange information is more operational than data from statistical services. But even based on operational data it is very difficult to predict processes, including the prices of goods. The value of an exchange commodity, say, crude oil, is determined not so much by the volume and cost of production or the level of supply and demand but rather by political decisions and decisions of stock exchange players, which usually have momentary and intuitive nature, which therefore cannot be measured by mathematical models. It is unrealistic to predict the price of crude oil using a point estimate or even a very small interval. The dynamics of the key factors affecting its price are undefined.

In any economy, where the rules of the game continually change, the probability assessment of a given event plays a primary role. It is even more important to assess the consequences of an event and consider that there may be unforeseen events, the probability of which cannot be estimated. For example, the change in the bank rate can be estimated with some degree of approximation, while a significant drop or rise in the prices of certain goods is almost impossible to predict. Thus, a scenario analysis should be present in any interval forecast. This approach is much more effective than a precise forecast.

For example, the exchange rate forecast can equalize all the outliers. This forecast can be presented as scientific only at first glance. In this case, provided such a forecast is agreed upon without any reservations, the economy ends up unprepared for the next unforeseen event. Even if the likelihood of this event is small, still the consequences may be disproportionate. Moreover, the consequences will certainly be nonlinear. At the same time, nonlinear assumptions often exacerbate the situation, as they create new qualitative difficulties for the interpretation of relations in the economy.

An attempt to identify causal links with the help of models can lead to neglect of the systemic nature of the economy and, hence, several key relations of the economy. At each step of a study, which is nonsystemic, any small assumption may result in a significant deviation from reality. The following analogy is appropriate here, and a similar chain can work in any model—black is almost the same as dark gray, dark gray is almost like gray, gray is almost like light gray and light gray is almost the same as white. Hence, black is almost the same as white. In this chain, qualitative changes occur, which at first glance are insignificant, but they lead to radically opposite results.

Small approximations at each step of a model structure can eventually result in a model that does not reflect reality at all. It is said that the devil is in the details, but if many details have their little devil, then in the fusion we get a big devil who can create all the conditions for destruction, rather than for creation.

Even a small change in the initial conditions can lead to a sharp increase in the likelihood of an event; if that happens, then the model through which the forecast is made is a fortiori unusable.[1]

A model will also fail if the process has passed its sensitivity threshold, that is, if the quality of the process has changed. If we rely only on model analysis, without considering economic patterns and the factor of uncertainty, then such an assessment of economic processes will be unacceptable.

Economists often try to squeeze out accurate forecasts. N. Silver's book quotes Jan Hatzius, chief economist at Goldman Sachs, who notes that there are three major problems for people who make economic forecasts. First, it is very difficult to identify causal relations based on an analysis of economic statistics alone. Second, the economy is constantly changing, so explanations of economic behavior that are appropriate for one business cycle may not be applicable for another. And third, not only are the economists' forecasts themselves bad but so are the data they work with. It is quite true that for forecasts, especially economic ones, where the data is filled with noise, statistical

---

1 Nassim Nicholas Taleb, *Antifragile: Things That Gain from Disorder* (New York: Random House, 2012).

calculations are more convincing if they are supported by theory or at least accompanied by deep thought about their bases.[2] Adjustments to the methods of statistical forecasts associated with certain judgments can improve their accuracy by 15 percent.[3]

A systemic study of patterns is necessary. An objective should be formulated as clearly as possible, and only then an attempt should be made to achieve it. The objective should not be adjusted to the convenient data or tools. If so, this puts it in a Procrustean bed of modeling, where adjustments, of course, are not a simplification but a serious misdeed.

§2. The Uncertainty Relations in Economy

The illusion of optimality

Uncertainty is an inherent feature of any economic system. Even if it is assumed that the economic relations are linear, the possibility of formulating the optimal criterion of the number of unknowns significantly exceeds the number of equations. After that, a part of the interrelated variables is subjectively declared as independent in the model. Effectively, they are defined outside the model. The other part is separated to be declared as dependent and calculated using those variables.

As we showed back in 1992 this does not match the economic reality.[4] The presented uncertainty relations of economic indicators emphasize that such a partition is conditional. Moreover, the extent of influence and counterinfluence of indicators are different. It changes not only in time but also in space.

The assumption that some indicators in a model depend on others and this dependence is constant is artificial. Moreover, not only the type of dependence is changing, but these changes are often impossible to foresee. The presentation of economic processes in the form of a set of variables and equations, where some variables determine the values of others, is conditional. The initial isolation of several variables as precisely definable is unacceptable because it immediately breaks economic relations. As Loretta J. Mester notes, "We might prefer to live in a world with more certainty, but we don't. And to pretend we do live in such a world is absurd—it can lead to bad outcomes."

---

2 N. Silver, *The Signal and the Noise: Why so Many Predictions Fail – But Some Don't* (New York: Penguin Press, 2012).

3 Stephen K. McNees, "The Role of Judgment in Macroeconomic Forecasting Accuracy," *International Journal of Forecasting* 6, no. 3 (October 1990): 287–99.

4 V. S. Dadayan and A. A. Tavadyan, *Systemology of Economic Indicators* (Moscow: Nauka, 1992).

The eighteenth-century French philosopher Voltaire noted, "Uncertainty is an uncomfortable position, but certainty is an absurd one."[5]

A determination of a few variables with a model fixes the others either clearly or to a certain degree.[6] Therefore, when choosing one group of variables in a certain way, the corresponding values of the other group of variables will be obtained.

With the preliminary fixation of one group of variables, the obtained values of other variables are not economic indicators in the full sense of the word, since their causal links are not fully considered. If the key economic indicators have a causal link, is it realistic to represent several indicators based on qualitative analysis and define others using mathematical methods? In a unified system of economic indicators, it is impossible to single out any one indicator from the general set for calculating others, since this breaks the causal links of indicators. In this case, some of the links will be unaccounted for. When dissecting some indicators from others, the elastic systemic relations of the economy are artificially presented as rigid causal relations. The dismemberment of a system and then its artificial reassembly are not acceptable and give bitter results in any field.

The relations of economic indicators can be represented using mathematical models. However, the most detailed record of economic links will not change the fundamental impossibility of determining the exact values of economic indicators. The conclusion that a group of key indicators only conditionally defines another represents the uncertainty relations of a system of economic indicators.

The causal links between economic indicators, implemented even by complex mathematical methods, are insufficient. The obtained results will be optimal for the model, but they may not be optimal for the real system of economic relations. The thing is, any mathematical model that makes it possible to represent the causal links between economic indicators does not depict the entire system of key relations. The assumption that economic processes can be determined using equations, where some indicators clearly define others, is artificial. This is because the equations do not consider the interdependencies, as well as the factor that the definition of causality in the economy, as a rule, is conditional. Besides, cause and effect can often transform and interchange.

---

5 Loretta J. Mester, *Acknowledging Uncertainty*, 10-07-2016; Shadow Open Market Committee Fall Meeting, New York, NY. Speech, Federal Reserve Bank of Cleveland Downloads https://econpapers.repec.org/paper/fipfedcsp/77.htm
6 The uncertainty relations of economic indicators are presented in the Appendix.

A model that reflects the true properties of an object is deterministic, even if it contains stochastic, probabilistic descriptions of causal links. This limitation of the model is unbiased not so much because of the model itself but because of the real properties of an economic system. The manifestation of uncertainty in economic reality lies in the fact that significant connections in the economy are precisely indeterminate, often ambiguous, and probabilistic and contain a tangible element of randomness, which is a natural property of an economic system.

To fully represent the system of economic interconnections, it is necessary to simultaneously determine the values of all key indicators. However, this objective cannot be realized. It is also impossible to obtain sufficiently accurate values of economic indicators because of the uncertainty relations.

The uncertainty relations thus presented do not reject the existence of causal links between economic indicators; it only asserts that there cannot be a precise determination of multiple economic indicators. It is impossible to determine the exact values of key indicators; it is only possible to specify their generalized relations.

The uncertainty relations allow us to assess the capabilities of the model. Optimal models are conditional, but in general terms, they show the relations of indicators. You should not reject a model out of the box if there is a rational grain in it. It is extremely important to indicate where and to what extent the model works because it is self-evident that you cannot throw useful things with the bathwater. It is much more reasonable to identify the real possibilities of a model, to define its limitations and to show where and how it can be useful and where it can go awry. For instance, models can help in formulating the most favorable interval of an indicator for economic development. However, the models do not make it feasibly possible to determine the exact values of an indicator. Note that the identification of the interval helps to form insurance measures if the economic indicator exits the said interval.

Mathematical optimization only makes it possible to formulate the uncertainty relations. Optimization can help to present the indicator connections in a generalized form, but by no means can it reveal the specific causal links in the economy even for the current situation, let alone its dynamics. This is the result of mathematical optimization, this is its purpose, and its possibilities should not be overstated or understated. Optimization models, like other models, are limited in the analysis of the real world. They do not describe economic processes in a unified system. They can only present the relations of economic indicators in a generalized way, as well as help to form an option on possible relations in the economy.

This can help to systemize the causal links of economic indicators, as well as in their interval estimation. Models can help to evaluate development options by significantly speeding up the computation process, although it should be borne in mind that with some probability those options will be in the uncertainty interval.

Nevertheless, the union of an applied specialist with a suitable model is a prerequisite for increasing the probability of successful forecasts but by no means enough for an accurate determination of economic indicators. It is unfeasible to precisely determine the dynamics of economic processes using mathematical models; however, the model can assess the relations of economic indicators in general terms, but no more.

Summary of Chapter 2

§1. The effect of compressed spring

- Equilibrium can be dangerous. With the accumulation of negative phenomena and with the transition of their quantity to a new quality, the "spring" of accumulated problems may be sharply released. Small doses of deviations, in contrast to rigid stability, reduce the risk of reaching the threshold of critical fluctuation. Slight stress allows us to identify the weak links in the system. A seemingly stable situation is by no means equivalent to a low-risk situation of negative consequences, especially in unforeseen circumstances.

- In any economy, where the rules of the game constantly change, the probability assessment of a given event plays a primary role. It is even more important to assess the consequences of an event and consider that there may be unforeseen events, the probability of which cannot be estimated. A scenario analysis should be present in any interval forecast. This approach is much more efficient than a precise forecast.

- A model will fail if the process has passed its sensitivity threshold, that is, if the quality of the process has changed. If we rely only on model analysis, without considering economic patterns and the factor of uncertainty, then such an assessment of economic processes will be unacceptable.

- A systemic study of patterns is necessary. An objective should be formulated as clearly as possible, and only then an attempt should be made to achieve it. The objective should not be adjusted to the convenient data or tools. If so, this puts it in a Procrustean bed of modeling, where adjustments are not a simplification but a serious misdeed.

§2. The uncertainty relations in economy

- Uncertainty is an inherent feature of any economic system. The assumption that some indicators in a model depend on others and this dependence is constant is artificial. Moreover, not only the type of dependence is changing, but these changes are often impossible to foresee. The presentation of economic processes in the form of a set of variables and equations is conditional. The initial isolation of several variables as precisely definable is unacceptable because it immediately breaks economic relations.

- The most detailed record of economic links will not change the fundamental impossibility of determining the exact values of economic indicators. The conclusion that a group of key indicators only conditionally defines another represents the uncertainty relations of a system of economic indicators.

- Obtained relations between economic indicators are insufficient. Those results will be optimal for the model, but they may not be optimal for the real system of economic relations. Any mathematical model that makes it possible to represent the causal links between economic indicators does not depict the entire system of key relations.

- To fully represent the system of economic interconnections, it is necessary to simultaneously determine the values of all key indicators. However, this objective cannot be realized. It is also impossible to obtain sufficiently accurate values of economic indicators because of the uncertainty relations. Their causal links can be presented only in a generalized form.

- Models can help to evaluate development options by significantly speeding up the computation process, although it should be borne in mind that with some probability those options will be in the uncertainty interval. It is unfeasible to precisely determine the dynamics of economic processes; however, the model can assess the relations of economic indicators in general terms, but no more.

Chapter 3

THE PRINCIPLE OF THE MINIMAL UNCERTAINTY INTERVAL

§1. The Minimal Uncertainty Intervals of Economic Indicators

To have an unattainable exact forecast or an achievable forecast in the minimal interval?

The interval uncertainty of key economic indicators represents the infeasibility to pinpoint an economic indicator within the minimal uncertainty interval. The probability of the minimal interval may change, and the interval will have to be adjusted. The minimal uncertainty interval should not be equated to the confidence interval; the uncertainty interval is not the mean value of an indicator with a statistical estimate of the probability of its deviation.

The presented uncertainty interval should not be regarded as a purely statistical tool with which it is possible, as in other intervals, to estimate the probability of a particular parameter of the population. Economic indicators are always changing, and their parameters and distributions also change over time. Not to mention the substantial amount of noise generated by the markets. Considering the factor of heterogeneity of economic indicators and in most cases small samples, the exact calculation of probabilities or p-values of one or another indicator is meaningless.

All outcomes are possible in the uncertainty interval; their probability is indefinable; a point estimate and the mean are an artifice in the interval. The mean or the average often gives the illusion of stability. The concept of the average is inefficient; the key here is the minimal uncertainty interval.

The average value is usually uninformative. The information can be obtained with the uncertainty interval and, of course, with an estimate of the likelihood of serious repercussions when exiting a said interval. Even if some value is declared as an average, the likelihood of its fulfillment does not become higher or more preferable than other possible values within the

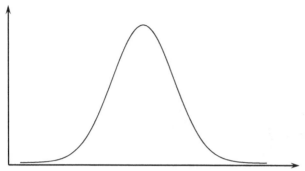

Figure 1 The curve of normal distribution.

interval. Indeed, this expression is appropriate here: do not jump off ledges in the dark if their average is a meter or three feet.

When inflation is expected to be between 2 and 5 percent, this does not mean that the likelihood of the average, which is 3.5 percent, will exceed 3 percent. For illustrative purposes, let's assume that the weather will change and the air temperature will drop by 4–6 degrees; this does not mean that the likelihood of 5 degrees is greater than 5.5 degrees. Moreover, the consequences of exiting the interval are quite different here.

In the sterile confidence interval, the margin of error is based on the most probable point, the sample means, which is inherently inaccurate and may not coincide with changing reality. Within the uncertainty interval, the probability of each point is indeterminate. It is not possible to calculate the distribution of values, say, like in the Gaussian curve (see Figure 1).

In the uncertainty interval, there is a high probability of obtaining the real values of economic indicators; however, it is impossible to precisely pinpoint the predicted value. The probability is much lower outside the uncertainty interval; however, the risks are greater. In this case, an economic indicator becomes much more vulnerable; harmful aftermaths are possible. Therefore, the value of the indicator can be assessed only within an interval. The aftermaths of the less likely outcomes of an indicator can significantly alter the causal relations in an economy. As such, the distribution of the predicted indicator has no single maximum point and entails sizable changes when exiting the uncertainty interval. Due to uncertainty, not only is the interval forecast of great value but also the assessment of the consequences when exiting the said interval.

Due to the dynamic nature of the economy, not only the high-probability but also low-probability critical values of an indicator are in the uncertainty interval. Economic indicators are more in line with the distribution shown in Figure 2.

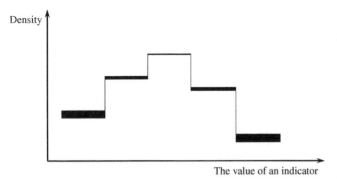

Figure 2 The most probable distribution of an economic indicator.

The marginal values of intervals are the estimated sensitivity thresholds of an economic indicator, at which the economic situation may significantly change. The width of the interval depicts the consequences for a given value.

Any values in an interval are equally likely within the discussed distribution. The situation is favorable in the most likely interval of an economic indicator. Decisions should be made considering this interval. The subsequent intervals are those in which changes in the situation can be evaluated and the relevant airbags for possible economic recessions can be developed. The least likely intervals may convey sizable economic shocks. Those should be considered if possible.

It is not possible to determine the probability of a value in the uncertainty interval, in contrast to the confidence interval, where it is theoretically assumed that the mean value has the highest probability.

The uncertainty intervals are estimated values. A minimal interval can be determined depending on the level of uncertainty. Notably when an interval of an indicator is determined too small, especially if it is a point estimate, then the probability of such a forecast is sharply reduced, becoming extremely small. A forecast with such a probability does not convey any value. In the other extreme, if a substantially large interval of possible values is considered, it also gives little information and, naturally, is also of little value.

In practice, during the 1990s, economists were able to predict only 2 out of 60 recessions in advance. Most point forecasts were presented, but signs of instability were not identified.[1] If we take the uncertainty interval for a GDP of 6.4 percentage points, then the forecasts will match the reality in

---

1 P. Loungani, "The Arcane Art of Predicting Recessions," *Financial Times via IMF*, December 18, 2000.

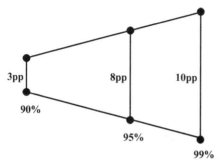

Figure 3 Acceptable minimal uncertainty intervals and the likelihood of their implementation.

90 percent of cases.[2] However, such a wide interval does not have practical value. Even with such a wide range of 6.4 percentage points, approximately 5 percent of recessions will fall out of it. However, when the interval is shortened, it is not easy to achieve the best result. Thus, the safety measures should always have a place in the forecast package.

Considering possible deviations, the construction of a much smaller interval would make it possible to develop scenarios inside the bounds and safety measures in case of exiting the interval. Achieving the real minimal interval of key indicators in the range of 3 percentage points with a 90 percent probability of its implementation can be considered a very successful result (see Figure 3).

The minimal interval should be of practical value. To illustrate this let's examine life expectancy given certain conditions. Extremes here are the exact estimates and ranges that encompass nearly all the possibilities of a life span.

Life Expectancy Forecast	*Description*
An exact figure, say, 77 years	Quackery
The minimal interval for determining the life expectancy	An acceptable result
0–100 years	Tautology

Nonetheless, given the challenges of forecasting, it is much more rational to expand the estimated uncertainty interval and to be prepared for unexpected situations than to engage in scientism. The uncertainty interval can be narrowed till its realization probability is not significantly reduced. This is beneficial for improving forecast efficiency. Besides, the probability itself is to

---

2 N. Silver, *The Signal and the Noise: Why so Many Predictions Fail – But Some Don't* (New York: Penguin Press, 2012).

be seen as an approximate value. Rigid forecasts do not contribute to the generation of safety airbags and the preparation of emergency measures in critical situations.

The difference between the lower and upper bounds of the key indicator is its minimal uncertainty interval for a given probability. Let's call this range the T-interval.

The fulfillment probability of an uncertainty interval depends not only on the adequacy of the said interval but also on the forecast horizon and the variability of the situation. This probability also depends on the dynamics of a specific economic indicator, other correlated indicators and how those correlations are detected and defined. Moreover, the quality of the initial data analysis and forecasting methods affect the uncertainty interval.

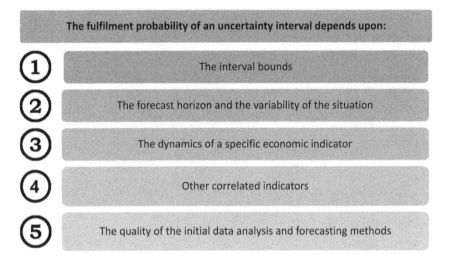

The fulfilment probability of an uncertainty interval depends upon:

1. The interval bounds
2. The forecast horizon and the variability of the situation
3. The dynamics of a specific economic indicator
4. Other correlated indicators
5. The quality of the initial data analysis and forecasting methods

The minimal uncertainty interval can be estimated depending on the above factors. However, the probability of a given value essentially cannot be determined within the interval. The more predictable the event, the smaller the uncertainty interval for a specific, usually short-term, forecast period.

A crisis presents so many new factors, the consequences of which cannot be estimated, that largely increases the uncertainty interval. This results in the need to present interval scenario forecasts with varying bands, which, if possible, should consider random events that directly affect both the interval and the reliability of the forecasts.

It should be noted that scenario forecasts can reduce the uncertainty interval of an economic indicator in each scenario. However, they cannot eliminate uncertainty by turning the interval into a point estimate. An overlap can be purely accidental, like a shot in the dark. Thus, decision making amounts to the determination of the most possible minimal uncertainty interval for key indicators and establishing scenarios for economic development with their implementation methods.

The diagnosis of the minimal interval of an economic indicator, including its extreme values, and possible consequences when obtaining such values is no less important than the forecast itself. When carrying out an interval forecast of an indicator, its diagnosis is also necessary. The utility of the forecast depends on the interval bounds—the less, the better. It also depends on how accurate the estimates of the possible consequences in case of obtaining results outside the interval are.

Possible, albeit unlikely, deviations from the uncertainty interval should be evaluated. If the consequences are significant when the economic indicator exits the uncertainty interval, then the economic system is fragile. In this case, particular outcomes can often be successful, but success will be unstable and short-lived. Let's say a monetary policy may be successful at first glance. However, if serious problems arise with the unlikely possibility of inflation or the national currency exiting their respected intervals, then such a monetary policy, of course, cannot be considered viable. If a study of the uncertainty interval does cover only mundane situations, without considering the probability of exiting the interval, which may have significant implications, there is a clear possibility of a serious setback, even a failure of the monetary policy.

The possible negative impact beyond the interval can be estimated as follows:

$$\text{Possible impact} = \int_{a}^{I_{lower}} \binom{\text{possible impact}}{\text{outside the lower bound}} + \int_{I_{upper}}^{b} \binom{\text{possible impact}}{\text{outside the upper bound}}$$

where I_{lower} and I_{upper} are the lower and upper bounds of an indicator, in the interval of which the probability of realization is the highest and a and b are the minimum and maximum possible values of an indicator.

If the possible negative impact does not increase significantly, it is feasible to narrow the uncertainty interval for the assessed indicator, because the impact of uncounted values is insignificant. However, this assessment is still preliminary, assessment of the bounds being quite subjective.

It could be right to shorten the forecast interval if there is no significant decrease in the fulfillment probability of the assessed interval. However, a precise forecast, especially in crises, significantly reduces the fulfillment probability and does not contribute to the creation of a mechanism for adjusting an economic policy. Doing a precise forecast is like trying to pinpoint what is ever-changing. In crises, the uncertainty interval sharply increases, and it is important to have several scenarios for the development of any economy, considering their implementation.

The method of the minimal uncertainty interval can cover the entire range of the most expected values. Of course, depending on the interval bounds of the key indicator, the corresponding airbags should be designed. Since it is impossible to achieve 100 percent realization of a given indicator in a reasonable interval, there are still rare events, some of which may have significant consequences.

The principle of minimal uncertainty interval gives forecasts the necessary flexibility and makes it possible to jump onto the last car of a train leaving for the future. The minimal intervals, which consider the economic patterns as well as its possible momentum and uncertainty for all future states of the economy, help to formulate the principles for forecasting economic processes.[3]

The basis for the most viable forecast is the minimal uncertainty interval. The forecast should proceed from the existing structure, development trends, and real possibilities of the economy. Economic forecasting should reside in the formation of forecasts in the minimal uncertainty intervals for key indicators.

Based on that concept, a system of intervals of key economic indicators will be laid down in this book. This is a necessary condition for economic efficiency.

§2. The Effect of the Expanding Uncertainty Bands in Economy

When forecasts lose their value

The presentation of programs for years ahead has little value because the uncertainty intervals expand when economic volatility increases, forming uncertainty bands. The rate of change in the world has increased; the economic situation has become highly volatile. It is essentially impossible to present sufficiently accurate long-term programs. It is advisable to formulate a direction for economic development and implement specific programs that improve the economic situation and contribute to economic development as soon as possible. It is also preferable to present possible scenarios.

Any economy is a dynamic, low-validity system, and the minimal intervals are acceptable only in the short run. The larger the span of the forecast, the

---

3 See Chapter 5 §2. Ten Principles for Forecasting Economic Processes.

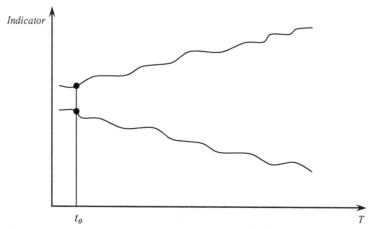

Figure 4 Uncertainty bands containing economic cycles.

wider the uncertainty interval of the most probable indicator values. The uncertainty band over time tends to expand significantly. Depending on the rate of expansion of the uncertainty band, forecasts become of little value; in fact, the formulation of its range after a certain time loses its practical meaning. Besides, for different indicators, the expansion process has its specifics.

Business cycles are in an uncertainty band. Moreover, the phase of cycles cannot be determined clearly. The bands, within which the business cycles are fitted, are presented over time in Figure 4.

The projections of business cycles are becoming progressively provisional because economic crises are increasingly unpredictable both in time and in depth. At point t_0, the forecast quality changes, and it becomes artificial. The economic indicator varies in an uncertainty band over time, wherein various scenarios are possible.

When unpredictability increases, the uncertainty band widens significantly while the time to reach that band is reduced. The significant change in the economic situation increases the likelihood of branching the uncertainty band of an economic indicator (see Figure 5). Quantity can change to a new quality. The likelihood of both upgrading (Scenario 2) and downgrading (Scenario 3) scenarios in an economy may increase.

Economic processes essentially have low validity; different and opposite scenarios may develop with wide variations of the uncertainty band. The further into the future, the larger the range of all possible variations of the uncertainty band. This is the nature of the economy and any low-validity system.

In a crisis, the uncertainty interval expands sharply, significantly increasing the range of the uncertainty band. Moreover, in a transforming economy, the band is prone to expand much more significantly.

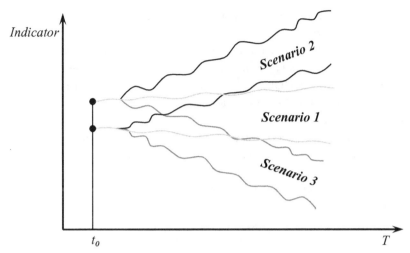

Figure 5 Possible scenarios of an uncertainty band.

§3. The Sensitivity Thresholds of Economy

The phase transition of elasticity, where the economic indicator changes dramatically

Any economic system has sensitivity thresholds, that is, the states that make it possible to reveal the critical bounds of key economic indicators, which allow suggesting their subsequent transition to a new quality. A comprehensive study of the sensitivity thresholds assists to reveal their numerical parameters, as well as the causal links of the economy. Given the increasing importance of the uncertainty factor in the economy, the estimate of a sensitivity threshold is only an approximation of the changing reality.

Sensitivity thresholds can have both positive and negative significance. The growth and the decline of GDP, exports and employment which bring the transition to the new economic quality describes positive or negative sensitivity thresholds respectively.

The systematization of sensitivity thresholds allows to determine the direction of transition of quantity to the new economic quality, which is most conducive to economic development. Achievements in economic policy will not lead to significant qualitative changes if the critical values of key indicators are not met. A change in any economic indicator does not automatically mean a qualitative change in the area it characterizes and much less that the sensitivity threshold has been achieved. The sensitivity threshold of an indicator occurs when the probability of transition to a new economic quality increases significantly in each area of the economy. In this case, achieving the sensitivity threshold of economic indicators is positive.

The reaction of economic, especially monetary, policy to the signals of disturbances in economic security is also essential. Those signals may include a sharp decline in GDP, an increase in the trade deficit and unemployment, a depreciation of the national currency and rising inflation. Near the sensitivity thresholds, even a minor change in an economic indicator could result in the negative transition of the economy to a new quality, which, for example, may entail a significant increase in prices.

In the conditions of uncertainty, the main factor of economic security is economic homeostasis, that is, steady development without fluctuations that pose a serious threat. The ability to withstand threats is an important condition for economic development. A slowdown in economic growth, especially a tangible recession, can lead to negative qualitative changes in the economy, which can have a long-term impact. In this case, sensitivity thresholds essentially are critical bounds beyond which the likelihood of unforeseen negative and substantial events increases significantly. Moreover, those bounds are dynamic and can change over time. To assess economic homeostasis, it is advisable to evaluate the thresholds which lead to a negative impact of quantity on the quality of economic growth.

The sensitivity threshold of an economic indicator outlines the possible consequences when exiting the bounds of the uncertainty interval. Beyond the sensitivity threshold, the elasticity of an economic indicator can undergo a phase transition because at this critical stage the indicator becomes very sensitive.

In an economy, the calculation of confidence intervals with certain probabilities does not matter much. Black swans can always appear out of the blue and change the scenery. What does matter is the evaluation of the values of indicators that may set off extreme events. Stepping out of those thresholds can cause economic phase transitions. The transition of sensitivity thresholds can change the nature and distribution of a given indicator. When passing a threshold, the indicator goes through a phase transition. For each indicator, those phase transitions have different characteristics. Those thresholds, however unlikely, have far-reaching repercussions, which can change the economic landscape, the causal links and the distribution of economic indicators.

The assessment of sensitivity thresholds for each economy is critical for economic development. The transition through those thresholds can cause both very negative and positive changes in the economy. Threshold transitions can help a transforming economy significantly improve the quality of life or bring economic downfall. That is why this book emphasizes what phase transitions of key economic indicators are and what they may entail.

It is difficult to pick a specific statistical tool that could calculate the sensitivity thresholds for each indicator. The thresholds are presented based on the historical experience of the development of countries and their transition to a new quality. Those thresholds are given for each key economic indicator and can serve as specific recipes for policymakers.

It is extremely important to identify the sensitivity thresholds, because the consequences beyond any threshold may be significant, even destructive. For oil and gas producing countries, where it is the main income, the fall in the market prices below a certain threshold may significantly reduce exports and GDP, devalue the currency, accelerate inflation and unemployment and significantly reduce the living standards.

When inflation goes beyond the sensitivity threshold, the level of profitability for economic entities becomes non-computable, with all the negative consequences for economic activity, including investments. If the influence of a key indicator is non-computable, that entails a sharp increase in the uncertainty interval of said key indicator.

An economic indicator can have several sensitivity thresholds. If this factor is not considered, then without a proper course of action the probability of critical situations increases significantly.

The depreciation of the national currency by 1 percent, 10 times during a year usually does much less harm than by 10 percent once a year. However, near the sensitivity threshold, the repercussions of 1 percent change may increase sharply, and the economic situation can abruptly alter. In unstable economic conditions, a 1 percent increase in the money supply may result in a transition of quantity to a new quality, that is, in critical situations, unforeseen negative sensitivity thresholds are formulated.

Economic indicators are prone to change, and the assessment of their possible sensitivity thresholds, the knowledge that they exist, of course, contributes to the forecast correction.

If the uncertainty interval is constructed so that the sensitivity threshold of the economic indicator is within its bounds, then significant changes may occur in that interval. Consequently, when constructing the uncertainty interval, it is advisable to analyze the possibility of finding the sensitivity threshold within it. It is also necessary to estimate the probability of the indicator going beyond the sensitivity threshold, being on the bound of the uncertainty interval. If this outcome has a high probability, it means that the possibility of both positive and negative substantial changes in the economy is significant.

Upon reaching the necessary condition for the transition of quantity into a new economic quality, which is the sensitivity threshold for an economic indicator, the possibility of substantial economic changes increases significantly. Without achieving the sensitivity thresholds that contribute to the development of an economy, it is impossible to obtain an acceptable level of stability, which ensures economic security and creates preconditions for the effective realization of economic potential.

The analysis of the critical characteristics of the key economic indicators—GDP, exports, trade balance, employment, inflation, bank rate, reserve requirement, national currency, level of monetization and foreign exchange reserves, public debt and the level of taxation—allows to lay down the parameters of the sensitivity thresholds of economic policy.[4] The sensitivity thresholds of the key economic indicators serve as benchmarks for an economy, which the economic measures must comply with.

Summary of Chapter 3

§1. The minimal uncertainty intervals of economic indicators

- A precise forecast, especially in crises, significantly reduces the fulfillment possibility and does not contribute to the creation of a mechanism for adjusting an economic policy. The probability of a given value essentially cannot be determined within the interval. Doing a precise forecast is like trying to pinpoint what is ever-changing. The construction of a substantially large interval of possible values gives little information and is also of little value.

- The uncertainty interval can be narrowed till its realization probability is not significantly reduced. This is beneficial for improving forecast efficiency. Besides, the probability itself is to be seen as an approximate value. Rigid forecasts do not contribute to the generation of safety airbags and the preparation of emergency measures in critical situations.

- The difference between the lower and upper bounds of the key indicator is its minimal uncertainty interval for a given probability. Let's call this range the T-interval.

- The minimal intervals, which consider the economic patterns as well as its possible momentum and uncertainty for all future states of the economy, help to formulate the principles for forecasting economic processes.

- The basis for the most viable forecast is the minimal uncertainty interval. The forecast should proceed from the existing structure. It could be right to shorten the forecast interval if there is no significant decrease in the fulfillment probability of the assessed interval. Decision making amounts to the determination of the most possible minimal uncertainty interval

---

4 See Chapter 4 §3. The Key Macroeconomic Task.

for key indicators and elaborating scenarios for economic development with their implementation methods.

§2. The effect of the expanding uncertainty bands in economy

- The presentation of programs for years ahead is of little value because the uncertainty intervals expand when economic volatility increases, forming uncertainty bands. It is advisable to formulate a direction for economic development and implement specific programs that improve the economic situation and contribute to economic development as soon as possible.

- Any economy is a dynamic, low-validity system, and the minimal intervals are acceptable only in the short run. The larger the span of the forecast, the wider the uncertainty interval of the most probable indicator values. The uncertainty band over time tends to expand significantly. Depending on the rate of expansion of the uncertainty band, forecasts become of little value; the formulation of its range after a certain time loses its practical meaning. Moreover, for different indicators, the expansion process has its specifics.

- Business cycles are in an uncertainty band, and the projections of business cycles are becoming progressively provisional because economic crises are increasingly unpredictable both in time and in depth. After some time, the forecast quality changes, and it becomes artificial. The economic indicator varies in an uncertainty band over time, wherein various scenarios are possible. Quantity can change to a new quality.

- Economic processes essentially have low validity; different and opposite scenarios of the uncertainty band may develop. The further into the future, the larger the range of all possible variations of the uncertainty band. This is the nature of the economy and any low validity system. In a crisis, the uncertainty interval expands sharply, significantly increasing the range of the uncertainty band.

§3. The sensitivity thresholds of economy

- The systematization of sensitivity thresholds allows to determine the direction of transition of quantity to the new economic quality, which is most conducive to economic development. A change in any economic indicator does not automatically mean a qualitative change in the area it characterizes and much less than the sensitivity threshold has been achieved. The sensitivity threshold of an indicator occurs when the probability of transition to a new economic quality increases significantly in each area of the economy.

- The main factor of economic security is economic homeostasis, that is, stable development without fluctuations that pose a serious threat. A slow-down in economic growth, especially a tangible recession, can result in negative qualitative changes in the economy, which can have a long-term impact. The identification of the sensitivity thresholds of the economy contributes to the assessment of economic homeostasis.

- The sensitivity threshold of an economic indicator outlines the possible consequences when exiting the bounds of the uncertainty interval. Beyond the sensitivity threshold, the elasticity of an economic indicator can undergo a phase transition.

- Without achieving the sensitivity thresholds that contribute to the development of an economy, it is impossible to obtain an acceptable level of stability, which ensures economic security and creates preconditions for the effective realization of the economic potential. The sensitivity thresholds of the key economic indicators serve as benchmarks for an economy with which the economic measures must comply.

Chapter 4

THE INTERVALS OF KEY ECONOMIC INDICATORS

§1. The Systematization of Economic Indicators

Conditions for stable economic development

Target indicators

Systematization of key indicators with the detection of their uncertainty intervals is a necessary condition for an efficient economic policy. Key indicators constitute nodes that link economic processes into one system. Links of key indicators accumulate other connections of economic processes and present them in a summarized form. The U.S. government estimates around 45,000 economic variables, and nongovernmental sources keep track of at least four million data.[1] In this case, many of the formulated variables will be of minor importance.

Thus, it is necessary to present the nodes, which summarize other variables to obtain tolerable results in the study of economic causations. This system approach should be carried out using a combination of economic objectives, normative constraints and the basic principles of regulation in the economy.

The system approach and a precise mating of key economic indicators are exceptionally important for economic development. Unfortunately, there are numerous examples of inconsistencies. For example, adherence to the principle of price stability often turns into a tough, inflexible monetary policy. Not enough attention is often paid to its feedback, that is, the explicit impact on GDP, structure of GDP, exports and employment.

In the most concentrated form, the key indicators are given and legislated in the so-called Magic square. In 1967, Germany adopted the Act to Promote Economic Stability and Growth, which formulated the basic principles of

---

1 Federal Reserve Economic Data. Economic Research, Federal Reserve Bank of St. Luis; Lakshman Achuthan and Anirvan Banerji, *Beating the Business Cycle: How to Predict and Profit from Turning Points in the Economy* (New York: Random House, 2004).

economic policy to avoid a subjective approach. Those principles for the key indicators, which are required as the basis for the preparation of budgets and financial planning, are as follows: steady economic growth, balanced foreign trade, a high level of employment and price stability.[2] The "Magic square" represents the key indicators of an economy in a concentrated form. They must be considered in a single system. An attempt to single out a sole indicator with no regard to its causations to other key indicators will not yield a quality result, especially in the long run.

Goodhart's law may snap into action if the monetary policy is represented by a single indicator.[3] Once a government or a central bank starts using a sole indicator, its significance may not match the economic reality. For example, if only minimal inflation is achieved at the expense of an excessively high bank rate and reserve requirement, reduction of foreign exchange reserves and monetization ratio, then this nonsystemic result is negative, and it does not contribute to economic development. The concentration on a single indicator makes the economic system volatile. Even if some GDP growth is recorded, this can neither ensure the stable development of an economy nor counteract crises.

For stable economic development, it is reasonable to present the primary economic objective in a synthesis, that is, exceeding the sensitivity threshold for GDP growth + a considerable increase in exports + a significant improvement of the balance of trade + employment growth. Those reference points are deliberately presented with a summation, for that is the way to successfully resolve economic problems holistically.

Substantial growth in exports is crucial, especially for the development of small economies, since the domestic market is limited by its nature. In developed countries with small economies, the share of exports of goods and services in GDP is greater than 50 percent, as it is clear that in a small economy the market is also limited; here exports become a decisive factor. This key indicator also needs a systemic approach. The coordination of economic programs, which are aimed not only at the growth of exports but also at achieving the exports to GDP ratio of 50 percent, becomes an important factor of the primary objective, especially in small economies. At the same time, it is necessary to modernize the structure of exports, determine the appropriate mechanisms to achieve this result and identify the threats and opportunities to counter them. Only when the exports to GDP ratio amounts to 50 percent, there will be a high probability of overcoming a critical bound—the sensitivity

---

2 Act to Promote Economic Stability and Growth. June 8, 1967, http://bit.ly/2FUajzK.

3 C. A. E. Goodhart, "Problems of Monetary Management: The U.K. Experience," Papers in Monetary Economics, Reserve Bank of Australia, 1975.

threshold for economic development. In this case, exports will fully contribute to substantial and qualitative changes in the economy.

Sustainable economic growth, balanced foreign trade and a high level of employment are certainly the basics for key target indicators in any economy. As for inflation, it is a key normative indicator, which has a causal link with target indicators. Price stability matching target indicators is a necessary condition for creating an opportunity to obtain the best results for target indicators.

Key normative indicators

Price stability that corresponds to the economic system is not the only necessary condition for economic development. In particular, the EU developed the Maastricht criteria, which are normative. Those criteria were consolidated in the Maastricht Treaty in 1992.[4] The following key indicators that are normative were highlighted: annual government deficit, government debt, inflation, exchange rate and long-term interest rates on government bonds. The bounds for those indicators are presented in the Maastricht Treaty.

Similar conditions to the Maastricht criteria have been implemented by other unions. For example, the Treaty on the Eurasian Economic Union formulates comparable indicators, concerning the specifics of the transforming economies. The Treaty presents requirements for regulatory key economic indicators that determine the stability of economic development.[5]

Table 1 presents the Maastricht criteria that ensure the balanced functioning of the economic and monetary union of the EU and the economic indicators that determine the stability of the economic development of the Eurasian Economic Union (EAEU).

The principles for coordinating the monetary policies in the EAEU are not quite developed. There is almost no specificity, which, of course, does not assist the definition of intervals for national currencies. The bounds for the exchange rate, which also has a regulatory function, are not determined by the Treaty.

It should be noted that both the Maastricht criteria and the criteria of key economic indicators presented in the Treaty on the EAEU display the objectives in a synthesis. Thus, even the key indicators presented in the Treaties should be considered with a certain conditionality.

Monetization ratio as well as the level of foreign exchange reserves, both directly related to inflation, should also be considered as normative indicators.

---

4 Convergence criteria—European Central Bank.
5 Treaty on the Eurasian Economic Union, Astana, 2015.

Table 1 EU Maastricht Criteria and Main Macroeconomic Indicators of the EAEU.

Indicator	EU	EAEU
Annual Government Deficit	The ratio of the annual government deficit to GDP must not exceed 3% at the end of the preceding financial year. If this is not the case, the ratio must have declined substantially and continuously and reached a level close to 3%.	The annual deficit of the consolidated budget of the government sector to GDP must not exceed 3%.
Government Debt	Government debt should not exceed 60% of GDP at the end of the fiscal year or strictly approach this level.	The debt of the government sector should not exceed 50% of GDP.
Inflation	The inflation rate of a given Member State must not exceed by more than 1.5 percentage points that of the 3 best-performing Member States in terms of price stability during the year preceding the examination of the situation in that Member State.	The annual inflation rate (December to December of the previous year) should not exceed 5 percentage points that of the lowest inflation in the Member States.
Exchange Rate	The observance of the normal fluctuation margins provided for by the exchange rate mechanism of the European Monetary System, for at least two years, without devaluing against the Euro.	The Treaty on the EAEU does not formulate specific conditions on the exchange rate.
Long-Term Interest Rate	The nominal long-term interest rate must not exceed by more than 2 percentage points that of, at most, the 3 best-performing Member States in terms of price stability.	There is no article regulating interest rates on government bonds.

§2. Intervals for Target and Regulatory Indicators

The minimum values of target indicators andacceptable values of regulatory indicators

Sensitivity thresholds for target indicators

The economic growth of less than 1 percent can be crucial for developed countries. Note that the level of GDP per capita is much higher in developed countries. In those countries, it significantly exceeds the average level of around $10,000. In terms of improving the living standards of developed countries, even a 1 percent GDP growth provides a tangible increase in GDP per capita. The approximate sensitivity threshold for GDP growth in developed countries that provides a sustainable increase of GDP per capita is as follows:

GDP growth rate ≥ 1 percent

Based on the experience of rapidly developing countries, to achieve substantial development in a transforming economy the annual GDP growth should be at least 5 percent. The benchmark to attain is the world average level of around $10,000 GDP per capita.

Below are the GDP growth intervals for transforming economies:

- GDP growth rate ≤ 0 percent—stagnation,
- 0 percent < GDP growth rate < 5 percent—quantitative economic growth,
- 5 percent ≤ GDP growth rate—an estimated sensitivity threshold for GDP in transforming economies, which creates conditions for qualitative growth.

Stagnation ≤ 0% < **Quantitative Economic Growth** < 5% ≤ **Qualitative Economic Growth**

For a considerable period, the average GDP growth rate in China, Singapore and South Korea has been well above 5 percent, as a result of transformations in economic regulation; thus the GDP per capita increased significantly there. China approached the level of $10,000 per capita in 2019, while the GDP growth rate over 10 years averaged 7.9 percent. Note that before 1999 GDP per capita was $3,800. Singapore achieved this level in 1994 and for 10 years the average GDP growth rate had been 9.3 percent. Here GDP per capita has only reached $2,500 before 1984. South Korea passed the level of $10,000 in 1989. Before that, the average GDP growth rate for 10 years had been 7.9 percent. Until 1979 GDP per capita did not exceed $5,000.[6]

As seen, in 10 years the average GDP growth rate in those countries significantly exceeded 5 percent. The growth rate of 5 percent gives a real chance of achieving this level, though over a longer period than 10 years.

Quite notable growth rates in China, Singapore and South Korea were accompanied by inflation that for mentioned periods averaged 2.5 percent

---

6 Data from World Economic Outlook Databases, IMF.

for China, 2.1 percent for Singapore and 5 percent for Korea. It is notable that during the process of German reunification in 1990–1992 inflation was 4 percent, with a comparably high average GDP growth rate of 3.8 percent.[7]

In countries where GDP per capita is well below \$10,000, significant improvements in living standards, including a noticeable pace of reaching this bound, do not occur if the annual growth rate is below 5 percent. With a below-average standard of living, a drop in GDP is especially substantial.

Sustainable GDP growth of 5 percent or more is possible if there is no significant negative trade balance and no deterioration of the balance of payments. For this, it is essential for small economies that the ratio of exports of goods and services to GDP exceeds 50 percent. Moreover, the number of finished goods that have high added value should prevail in exports.

Exports to GDP ≥ 50%

In countries where the population is less than 10 million and GDP per capita is well above the average, the exports to GDP ratio has been 56 percent for Austria, 66 percent for Switzerland, 56 percent for Denmark, 176 percent for Singapore and 122 percent for Ireland. In those countries, the export to import ratio has been 107 percent for Austria, 122 percent for Switzerland, 112 percent for Denmark, 117 percent for Singapore and 112 percent for Ireland. Employment has been 95 percent for Austria, 95 percent for Switzerland, 95 percent for Denmark, 96 percent for Singapore and 94 percent for Ireland. Note that in countries with a per capita income above the average, the monetization ratio is well over 50 percent. For instance, in Denmark, the monetization ratio has been 61 percent, and in Switzerland and Singapore, it even exceeds 100 percent—189 percent and 123 percent, respectively.

The probability of achieving the exports to GDP ratio of 50 percent increases significantly if the trade balance, the difference between exports and imports, is not negative;

Balance of Trade ≥ 0
meanwhile, employment should be at least 90 percent.

Employment ≥ 90%

---

7 World Economic Outlook Databases, IMF.

Thus, the sensitivity thresholds for target indicators that any country should strive to achieve are systematized below:

- **GDP Growth**   ≥ **1%**   for developed countries
                    ≥ **5%**   for transforming economies

- **Exports to GDP**   ≥ **50%**   for small economies

- **Balance of Trade**   ≥ **0**

- **Employment**   ≥ **90%**

Those bounds represent sensitivity thresholds for key target indicators. If those target conditions are not met, the transition of the sensitivity threshold of economic development, which ensures the qualitative changes in the economy, is unlikely.

Changes in the uncertainty interval of one key indicator may alter the interval of another indicator because they are bound in a single system. Thus, normative and regulatory indicators, as well as target indicators—should be estimated in intervals. Such a problem, certainly, cannot have a point solution.

Forecasts are highly dependent on economic policy. If normative or regulatory indicators exit their favorable and acceptable intervals, this may result in a significant deviation of target indicators from their forecast value. Thus, normative and regulatory indicators should remain in their favorable and acceptable intervals, respectively.

Intervals for normative indicators

If for developed countries inflation is favorable in the interval of 0 percent ≤ inflation ≤ 2 percent, then for transforming economies inflation is appropriate in the interval of 2 percent ≤ inflation ≤ 5 percent.

The monetization ratio in developed countries is higher and exceeds the required sensitivity threshold, while exchange rate fluctuations are normally insignificant. The policy of lowest inflation is acceptable only if the balance of trade is positive, the balance of payments does not deteriorate and the monetization ratio corresponds to the favorable changes in the economy.

Let's stress that the economic policies of developed countries should not be equated to the policies of transforming economies. The practice of literal duplication of economic policies of developed countries in any transforming economy is irrational, and the result is unlikely to be comforting. Imagine an attempt by a rookie sportsman to perform exercises developed for a prime athlete. The result is unlikely to be reassuring, and injury is possible.

Here is the favorable inflation interval for developed countries

0 percent ≤ Inflation ≤ 2 percent
and for transforming economies
2 percent ≤ Inflation ≤ 5percent.

Note that the upper bound of the interval is 2 percent for developed countries and 5 percent for transforming economies. This bound is influenced by the dynamics and structure of inflation, especially by the rise in prices of essential products. If annual inflation remains in those favorable bounds but has significant price fluctuations, this may cause economic instability, particularly give a notable jitter to the exchange rate.

Inflation that corresponds to the real state of the economy will gradually decrease to the level acceptable for developed countries. Above all this ensures its long-term stability. Only the strategy that considers the real causal links of key economic indicators is favorable for assuring positive changes in an economy.

Outlined intervals do facilitate economic growth, for the most possible low inflation should meet the following criteria: it must not deteriorate the balance of trade and payments, assist unemployment, should not result in lowering monetization ratio and foreign exchange reserves.

Here is the central dilemma. Is it better for the development of a transforming economy to have the lowest possible inflation that does not coincide with the aforementioned conditions or relatively low inflation that does not exceed a certain bound? To choose between seemingly low inflation, which is not linked to other key indicators, may have seasonal and structural fluctuations, or inflation which has minimal fluctuations that do not exceed the acceptable bound and correspond to other key indicators. The choice is obvious here.

Monetary policy that is coordinated with fiscal policy is preferable, rather than out-of-system price stability. Policy coordination stimulates economic growth and contributes to financial and price stability. A central bank is indeed independent in the implementation of monetary policy, although it must act within a single economic system.

Considering the systemic approach to price stability, the monetization ratio should meet the following conditions when inflation is in a favorable interval:

Monetization Ratio   ≥ 50%   minimal sensitivity threshold
                     ≥ 70%   favorable sensitivity threshold

A monetization ratio exceeding 50 percent is a necessary condition for stable economic growth; a monetization ratio exceeding 70 percent is a favorable sensitivity threshold for economic development. Otherwise, the monetization ratio will not be sufficient and favorable for investment, development and promotion of new technologies; significant development of infrastructure; and production of new goods and services.

Below is the optimal interval of foreign exchange reserves:

25 percent of Imports ≤ Foreign exchange reserves ≤ 100 percent of Imports

The IMF as a traditional rule of thumb states that reserves are adequate if they cover three months' worth of imports, especially for countries where imports are a significant expense article.[8] This condition is particularly important for countries where the ratio of imports to GDP exceeds 50 percent. If foreign exchange reserves cross the lower bound, this will dramatically reduce the potential for regulation through foreign exchange reserves. The possibility of a positive impact on financial stability through this normative indicator is significantly reduced. If foreign exchange reserves exceed annual imports, then this can considerably decrease the growth potential of the economy, because the growth of foreign exchange reserves may occur at the expense of economic growth.

According to the Maastricht Treaty, government debt should not exceed 60 percent of GDP at the end of a fiscal year or steadily approach this level. In the EAEU, this condition is even stricter: the government debt should not exceed 50 percent of GDP. Taking into account that the economic system is less resistant to unexpected shocks, especially in transforming economies, and that external debt usually has a high percentage in government debt, this limitation is more appropriate for such countries.

Government Debt   ≤ 60%   Sensitivity threshold for developed countries
                 ≤ 50%   Sensitivity threshold for transforming economies

Below is the bound for government deficit. The same criteria for this indicator are implemented both in the EU Maastricht criteria and the Treaty on the EAEU.

Government Deficit ≤ 3 percent

The interest rate on government bonds is guided by the bank rate, which is a key regulatory indicator. The value of the national currency, its exchange

---

8 IMF Survey: Assessing the Need for Foreign Currency Reserves.

rate, which is directly related to the level of inflation, is also of a regulatory nature. Given the permissible intervals of regulatory indicators, the comprehensive implementation of the selected restrictions for key economic indicators is a complicated task of regulating the economic processes.

§3. The Key Macroeconomic Task

The art of regulation within the intervals for key economic indicators

The key indicators for all developed, rapidly developing and transforming economies are similar. Target indicators are described by intervals of economic development, normative and regulatory indicators, by favorable and acceptable intervals, respectively.

Monetary indicators react faster to changes in an economic system. They have more elasticity than fiscal indicators, which normally require some necessary legislative procedures and time to change.

Monetary policy plays a leading role, especially when mitigating the negative aftermaths of economic shifts, because it allows a quick and efficient response to economic shocks.

Intervals for bank rate

Bank or discount rate plays a leading role in the regulation of inflation, although reserve requirement, national currency and the transactions of government bonds also play a certain role.

Let's present the lower and upper interval bounds of the bank rate. Inflation has a key part in the lower bound of this interval. The bank rate should not usually be less than the inflation. Since inflation changes depending on the economic situation, the benchmark for the lower bound of the bank rate also varies.

The benchmark for the upper interval bound of the bank rate is evaluated by subtracting the holding period return for areas that ensure scientific and technological progress to the net interest margin. This plays a decisive role for the upper bound of the real bank rate (RBR, the difference between the bank rate and inflation), which, as shown by the experience of countries in periods of significant economic growth, cannot be more than 2 percent.

The value of RBR that is more than 2 percent can only be justified in critical situations, especially in conditions of low monetization ratio. Moreover, the deviation can be validated only in a short term, as it may cause serious damage to the economic potential of the country.

Considering the sensitivity threshold, let us assess the admissible minimal and maximal real bank rates. The minimal bank rate is the lower sensitivity threshold, which may cause higher inflation. The maximal is the upper sensitivity threshold that may hinder economic growth.

Stable economic growth is unlikely if the real bank rate, the difference between the bank rate and inflation, is more than 2 percentage points. This pattern of economic growth is also confirmed in practice.

There is not any huge difference between the bank rate and inflation in developed countries or in countries with a significant rate of economic development. The real bank rate in those countries is not only below 2 percent, but it even approaches 0.

The experience of the United States, Canada, the EU countries and rapidly developing countries shows that low inflation is ensured with real bank rate not exceeding 2 percent, and often this rate is even smaller. Monetary policy will not ensure economic development by lowering inflation to its minimal possible bound. The lowest possible inflation does not play a key role in monetary policy to ensure economic development. Developed and rapidly developing countries achieve this bound of low inflation gradually. Economic development for any type of economy is in essence ensured if the difference between the bank rate and inflation does not exceed 2 percentage points. This measure promotes an easy money policy.

In the countries listed above, the difference between the bank rate and inflation does not exceed 1 percentage point. On the other hand, in countries with transforming economies, even with low inflation, the difference between the bank rate and inflation often significantly exceeds 2 percentage points, which, of course, does not contribute to economic development. Moreover, the nonmonetary component of inflation in transforming economies is significant; thus, a high bank rate is usually not effective when regulating inflation in those economies.

If low inflation is achieved by the means of unfavorable real bank rate, high reserve requirement and overvalued currency, this inevitably will hurt the economic development. As a rule, in those conditions, favorable terms for investments and exports are sacrificed, the balance of payments deteriorates and foreign exchange reserves and the monetization ratio decrease.

Therefore, the acceptable interval for the real bank rate is as follows:

$$0 \text{ percent} \leq \text{Real Bank Rate} \leq 2 \text{ percent}.$$

To ensure economic development the difference between the bank rate and inflation should not be more than 2 percentage points. This value is most consistent with the effective coordination of monetary and fiscal policies. Namely, this interval most of all contributes to GDP and exports growth, especially to the growth of exports in finished goods, the formation of the corresponding exchange rate, while it has a positive effect on the growth of monetization ratio.

In this context, it is necessary to ensure the inflation for developed countries in the interval of 0–2 percent and for the transforming economies 2–5 percent. This can be considered a key objective of monetary policy.

The closer the real bank rate is to the lower bound, that is 0 percent, the more effectively the economic system functions, if, of course, the lower bound of the real bank rate allows to restrain inflation for the transforming economy below 5 percent and for the developed economy below 2 percent.

It is the real bank rate, which corresponds to the economic development, that makes it possible to gradually reduce inflation and contributes to long-term stability. Only a strategy that considers real causations is favorable for positive changes in the economy.

A transforming economy frequently employs unreasonably high bank rates as well as a high reserve requirement if the inflation is above 5 percent and, paradoxically, below 2 percent. Moreover, especially in the second case, the real bank rate may be extremely high. If so, measures are taken to limit the monetization ratio well below its sensitivity threshold to ensure inflation under 2 percent.

A transforming economy is likely to implement an unreasonably high real bank rate when it attempts to keep inflation at an extremely low level. This is accompanied by an exchange rate that seems to be extremely stable. This is especially inefficient if inflation is imported.[9] The economic penalty for this is not only a reduction of monetization ratio and foreign exchange reserves but most importantly the decline in the balance of payments and the balance of trade. This will inevitably reduce the rate of economic development. The probability of those negative consequences is significantly reduced if the real bank rate and exchange rate are corresponding to the current economic system.

Monetary policy as an integral part of economic policy should be implemented in a systemic and unified manner. The reduction of bank rates has a positive effect on the process of coordinating monetary and fiscal policies. This measure also has a positive effect on the gradual increase of monetization ratio, bringing it closer to the favorable sensitivity threshold for countries with transforming economies. The monetary target is not to ensure the lowest inflation by all possible means but to have inflation corresponding to the reality, which above all contributes to the growth of the economy and exports

---

9 Joseph E. Stiglitz, *The Price of Inequality: How Today's Divided Society Endangers Our Future* (New York: W.W. Norton & Co, 2012).

and at the same time has a positive effect on monetization ratio, bringing it closer to its sensitivity threshold.

A high interest rate may be a motivation for an organization seeking the highest revenue, but it certainly cannot be a motivation for any central bank. Rapid economic development cannot occur in a country that has a high price for extremely low inflation. Experience shows that in such conditions an unfavorable situation is created for economic development, regarding investments, exports and foreign exchange reserves.

Bringing the real bank rate in line with the tasks of economic development and gradually reducing it to 2 percent is a necessary condition for economic growth. This measure creates opportunities for stable low inflation and significantly reduces the likelihood of notable fluctuations in the exchange rate. Note that if the surplus money is used for investments, it does not hurt the real bank rate. In this case, there is no need to raise the bank rate to curb inflation. Of course, a reduction in the difference between the bank rate and inflation, as well as a reduction in the reserve requirement, may lead to currency depreciation and, accordingly, to a certain increase in prices for imported goods. However, such a monetary policy has a positive effect upon the balance of payments, the balance of trade and as a whole on the country's economic development.

An important goal is to prevent inflation from rising above its acceptable upper limit while adhering to the downward trend for the real bank rate. Given the fact that in transforming economies inflation is often nonmonetary, naturally, it is directly affected by the rise in prices of imported goods.

The following causal chain may have a decisive role, especially considering the substantial nonmonetary component of inflation in transforming economies: excessively high real bank rate → reduction of monetization ratio → high-interest rate → cost-push inflation. The increase in monetization ratio is a positive consequence of the low real bank rate, as well as the low reserve requirement. Note that the monetization ratio is an indicator of the effectiveness of the monetary policy.

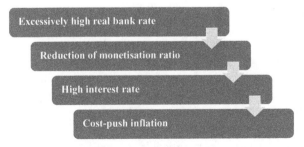

The effect on the economy when the real bank rate is above the acceptable threshold

The following rule derives from the properties of the real bank rate. If for a long period the real bank rate is greater than 2 percent, especially when the monetization ratio is less than 50 percent, this results in expensive money, which is an obstacle to economic growth. This tendency can be amplified by the growth of the reserve requirement, wrong steps in currency regulation and off-target transactions of government bonds.

If 0–2 percent inflation is acceptable for developed countries, then the permissible nominal bank rate in these countries should be within the interval of 0–4 percent.

0% ≤	**Nominal Bank Rate**	**≤ 4%**
the minimum at 0% inflation	**(Developed Countries)**	the maximum at 2% inflation

For transforming economies, where inflation is acceptable in the interval of 2–5 percent, the nominal bank rate should be in the following interval:

2% ≤	**Nominal Bank Rate**	**≤ 7%**
the minimum at 2%inflation	**(Transforming Economies)**	the maximum at 5%inflation

In effect, the intervals of 4 and 5 percentage points (4–0 percent and 7–2 percent) are the benchmarks for nominal bank rate regulation in developed countries and in transforming economies, respectively.

The acceptable interval for the reserve requirement is from 0 percent to 10 percent. The sensitivity threshold for inflation regulation is 0 percent and the sensitivity threshold for business is 10 percent.

$$0\% \leq \textbf{Reserve Requirement} \leq 10\%$$

Intervals for exchange rate

Currency depreciation which follows a gradual and steady trend has a positive effect on economic growth, primarily on the growth of exports. Gradual currency depreciation is also beneficial for the implementation of the state budget. However, it should be emphasized that even a gradual currency depreciation cannot be an end in itself. The currency may be affected by a decrease in the real bank rate, but at the same time, the rate of the national currency should be within its permissible interval.

The decrease in the real bank rate contributes to an increase of monetization ratio, under the condition that inflation does not exceed the acceptable upper bound and does not have significant fluctuations within the interval, which is also important. If in parallel the national currency steadily depreciates within its permissible interval, then this cannot be considered a negative phenomenon. If a seemingly stable exchange rate disagrees with economic development, the probability of a sharp and sudden leap of the exchange rate becomes extremely high.

The following example is illustrative. Is it safer to jump from half a meter 24 times a year or have a high probability of falling from 10 meters high? It seems in the first case the cumulative jumps amount to a greater height of 12 meters, and in the second case, there is only some probability of falling and from a lower height. However, the consequences are incomparable; in the first case it is practically safe, and in the second there are certainly serious, even grave consequences. In addition, the damage grows nonlinearly, meaning it will increase not only 20 times (10/0.5) but many more. It is sensible that it is preferable to have periodic currency fluctuations of 0.5–1 percent to ensure stable economic growth, rather than to have a high probability of a sharp depreciation of 10 percent or more, which happened to the currencies of the EAEU countries during 2014–2015. In September 2011 this also happened to the Swiss franc, which is declared to be a pegged currency.[10]

Significant national currency depreciation, first of all, harms the modernization of industries using imported equipment. National currency depreciation may spur inflation, especially if imports significantly exceed exports. On the other hand, significant national currency appreciation is even more unfavorable and poses a clear threat of decline in exports and economic growth.

Currency depreciation in general is favorable for export growth. However, due to the national currency depreciation, the price of imported raw materials and equipment may increase significantly. This will notably reduce the profitability of those goods. Hence, the demand for imported goods may decrease. On the other hand, if the national currency appreciates excessively, the price of domestic products in external markets will increase significantly, which will have a direct negative effect on its competitiveness.

The permissible interval of the exchange rate forms its sensitivity thresholds for exports and imports, accordingly. Exceeding the sensitivity threshold of the exchange rate poses a threat to exports and economic growth. The

---

10 Kaushik Basu and Aristomene Varoudakis, *How to Move the Exchange Rate If You Must: The Diverse Practice of Foreign Exchange Intervention by Central Banks and a Proposal for Doing It Better* (The World Bank, May 2013), Policy Research Working Paper; No. 6460. Washington, DC.

depreciation of the national currency below the sensitivity threshold, first of all, poses a threat for production with imported raw materials and equipment.

Hence, the exchange rate contributes to economic development and does not deteriorate the balance of payments and the balance of trade if it is in the interval presented below:

$\underline{\$} \leq$	**Exchange Rate**	$\leq \overline{\$}$
The sensitivity threshold for imports. Threat to production with imported raw materials and equipment.		The sensitivity threshold for exports. The danger for exports and economic growth.

where $\overline{\$}$ is the sensitivity threshold of the exchange rate for exports, when further national currency appreciation is a threat to exports and economic growth. $\underline{\$}$ is the sensitivity threshold of the exchange rate for imports, when further depreciation is a threat to industries using imported raw materials and equipment.

Currency can have a dampening function. It helps to absorb internal and external economic shocks. Small fluctuations of the national currency within its permissible interval are preferable to a rigidly stable rate with a high risk of a sharp departure from its permissible interval. The paradox is as follows: for the stability of the currency, it must display small fluctuations. Currency fluctuates to remain stable, rather than to leap.

For the floating exchange rate, its interval bounds indicate the feasibility of the regulatory steps. With the primary objective to significantly increase exports, which is extremely important especially for small economies, it is advisable to promote the formation of the exchange rate closer to its possible lower threshold, but this bound should not be crossed. If crossed this can not only restrict the imports of equipment and raw materials but also provoke inflation above its favorable interval, contributing to the transition of its sensitivity threshold.

Intervals for taxation

The key purpose of taxes is to ensure the solution of priority social problems, the needs of state administration, the proper defense level and the development of fundamental routes of scientific and technological progress. This determines the minimal amount of taxation—the threshold of taxation sensitivity for the state. This is primarily ensured by indirect taxes, value-added tax and turnover tax.

In addition to the fiscal and social functions, taxation also has a regulatory function. An important role of taxation is the contribution to the economic growth, especially to the exports of finished goods, through its impact on prices and, hence, on the volume of output. Fiscal indicators in conjunction with monetary indicators are the key levers of economic regulation; however, it should be borne in mind that their impact should be consistent with the acceptable intervals of key indicators.

The level of taxation should not lead to a decrease in GDP growth rates and, hence, to the reduction of the output and the respective rise in prices. Otherwise, there may come about processes that would hinder economic development. In this case, the following causal chain will act: taxes exceeding the permissible threshold → high expenses → low profitability and reduction in output → a decrease in possible GDP growth, including exports → a decrease in the country's competitiveness.

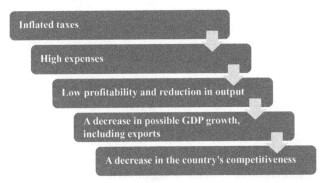

The effect on the economy when the taxation level is above the acceptable threshold

The lower and upper thresholds of taxation determine the minimum level of taxation required to perform the tasks of the state and the maximum level which does not affect the economic development, respectively.

Based on the volatility of the economic situation, the possibility of making rational decisions is determined by the following intervals for taxation:

$$\underline{T} \leq \textbf{Level of taxation} \leq \overline{T}$$

where $\underline{T}$ is the lower threshold of taxation, ensuring the minimal performance of the state, and $\overline{T}$ is the upper threshold of taxation that does not impede economic development.

It should be noted that the sensitivity thresholds for taxation depend on the efficiency of government expenditures.

The key macroeconomic task

The key macroeconomic task resides in the harmonization of monetary and fiscal policy, which is aimed at economic growth, primarily at the growth of exports and employment. Moreover, the restrictions on normative indicators must be met, and the regulatory indicators must be within their acceptable bounds.

For target indicators, the aim is to maximize their value. Their sensitivity thresholds are as follows:

- GDP growth rate ≥ 1 percent, for developed countries
  GDP growth rate ≥ 5 percent, for transforming economies
- Exports of goods and services to GDP ≥ 50 percent, for small economies
- Balance of trade ≥ 0 percent
- Employment ≥ 90 percent

Bounds on normative indicators:

- 0 percent $\leq$ Inflation ≤ 2 percent, for developed countries
  2 percent $\leq$ Inflation ≤ 5 percent, for transforming economies
- Government debt to GDP ≤ 60 percent, for developed countries
  Government debt to GDP ≤ 50 percent, for transforming economies
- Government deficit to GDP ≤ 3 percent
- Monetization ratio ≥ 50 percent minimal threshold
  Monetization ratio ≥ 70 percent favorable threshold
- 25 percent of Imports $\leq$ Foreign exchange reserves ≤ 100 percent of Imports

The acceptable intervals of the regulatory indicators:

- 0 percent $\leq$ Bank Rate ≤ 4 percent, for developed countries
  4 percent $\leq$ Bank Rate ≤ 7 percent, for transforming economies
- 0 percent $\leq$ Real Bank Rate ≤ 2 percent
- 0 percent $\leq$ Reserve Requirement ≤ 10 percent
- $\underline{\$} \leq$ Exchange Rate $\leq \overline{\$}$
- $\underline{T} \leq$ Level of Taxation $\leq \overline{T}$

The key macroeconomic task makes it possible to overcome the complexity of the systemic representation of the terms ensuring sustainable economic development.

The game rules in economy

Altering the game rules in the economy can be crucial, and shifts in any indicator can significantly transform economic processes. To avoid negative alterations the relevant changes should occur for normative indicators in their favorable intervals and regulatory indicators in their acceptable intervals.

It should be noted that forecasts provided by different international organizations often differ significantly and are adjusted every three months; after the adjustment, they usually remain unchanged for another three months. Economic forecasts have little value without the assessment of possible deviations and their consequences. In the macroeconomic task thus outlined, it is rather substantial to assess the probability of favorable and acceptable intervals and sensitivity thresholds for normative and regulatory indicators, respectively. Favorable and acceptable intervals of key indicators are in fact guidelines for their minimal uncertainty interval. It is reasonable to make sure that they remain in the specified intervals of the macroeconomic task with high probability.

It is not easy to imagine a change in the rule book mid-game. During sporting events, changes to the game rules are unheard of, but they are possible in the economy. The terms in the economy are constantly changing and evolving. These continuous processes directly affect the predicted values of indicators aimed for economic development. As a result, normative and regulatory indicators must be constantly adjusted to the new conditions, based on subjective and objective opportunities. This is the essence of the real economy.

Political and economic uncertainty leads to the fact that in the best case it is possible to predict the key target indicators characterizing economic development in specific intervals. Those target indicators are the GDP, exports, trade balance and employment. In particular, the forecast quality is affected by fluctuations in prices for mineral products, mainly for energy resources, which affect economic growth in many countries; changes in the bank rate of the leading countries and unions with the largest share of GDP in the world; and the pace of economic growth in those countries wherein products are exported.

The art of economic regulation

The art of economic regulation lies in the determination of acceptable intervals for regulatory indicators to overcome the positive sensitivity thresholds of target indicators, given the normative indicators remain in their favorable intervals.

Economic growth tends to become random and unstable without considering the limitations of normative indicators and acceptable intervals of regulatory indicators, while the adjustment of regulatory indicators in their acceptable intervals promotes the development of the economy and stabilizes the situation.

The formation of a system of favorable and acceptable intervals for key normative and regulatory indicators, respectively, can substantially reduce the uncertainty interval of target economic indicators. Minimal intervals of normative and regulatory indicators help to predict possible values of target indicators.

The difference between the upper and lower bounds of regulatory indicators determines the scope of the art of regulation influencing the economic processes. The art of regulation is defined as finding an effective solution within this scope.

The necessary conditions for an efficient economic system are as follows:

1. The determination of possible intervals of a key target, normative and regulatory indicators;
2. The systematization of key indicators, their presentation in the form of a single macroeconomic task;
3. The identification of possible methods of regulating the economic indicators, including the inefficient ones;
4. The elaboration of rational methods of regulating economy under uncertainty;
5. The adoption of efficient solutions, matching the conditions of economic performance, as well as their evaluation.

Summary of Chapter 4

§1. The systematization of economic indicators

- It is necessary to systematize key indicators and identify their uncertainty interval to obtain acceptable results in the study of economic causations. This systemic approach should be carried out in a combination of economic objectives, normative constraints and the basic principles of regulation in the economy. Key indicators constitute nodes that link economic processes into a single system. Links of key indicators accumulate other connections of economic processes and present them in a summarized form.

- An attempt to single out a single indicator with no regard to its causations to other key indicators will not yield a quality result, especially in the long run. The concentration on just one indicator makes the economic system volatile. Even if certain GDP growth is registered, this can neither ensure the stable development of an economy nor counteract crises.

- For stable economic development, it is reasonable to present the primary economic objective in a synthesis, that is, exceeding the sensitivity threshold for GDP growth + a considerable increase in exports + a significant improvement of balance of trade + employment growth. Those reference points are presented with a summation, for this is the way to successfully resolve economic problems holistically.

- To modernize the structure of exports it is necessary to determine the appropriate mechanisms for achieving this result as well as for identifying the threats and capabilities to counter them. In small economies only when the exports to GDP ratio amounts to 50 percent, there will be a high probability of overcoming the sensitivity threshold for economic development. In this case, exports will fully contribute to substantial and qualitative changes in the economy.

§2. Intervals for target and regulatory indicators

- Achieving the sensitivity thresholds of target indicators is crucial for stable economic development. If the sensitivity threshold is not reached for a target indicator, then qualitative changes in the economy become unlikely.

- Forecasts are highly dependent on economic policy. If normative or regulatory indicators exit their acceptable and favorable intervals, this may result in a significant deviation of target indicators from their forecast value.

- Monetary policy that is coordinated with fiscal policy is preferable, rather than out-of-system price stability. Policy coordination stimulates economic growth and contributes to financial and price stability. A central bank is independent in the implementation of monetary policy, although it must act within a single economic system. Only a strategy that considers real causations is favorable for positive changes in the economy.

§3. The key macroeconomic task

- Monetary indicators react faster to changes in an economic system. They have more elasticity unlike fiscal indicators, which normally require some necessary legislative procedures and time to change. Monetary policy plays a leading role, especially when mitigating the negative aftermaths of economic shifts, because it allows a quick and effective response to economic shocks.

- To ensure economic development the difference between the bank rate and inflation should not be more than 2 percentage points, which is most consistent with the effective coordination of monetary and fiscal policies.
- The real bank rate of no more than 2 percent coincides the most with GDP and exports growth, especially to the growth of exports in finished goods, the formation of the corresponding exchange rate, while having a positive effect on the growth of monetization ratio. In this context, it is necessary to ensure the inflation for developed countries in the interval of 0–2 percent and for the transforming economies 2–5 percent. This can be considered a key objective of monetary policy.
- The permissible interval of the exchange rate forms its sensitivity thresholds for exports and imports, accordingly. Exceeding the sensitivity threshold of the exchange rate poses a threat to exports and economic growth. The depreciation of the national currency below the sensitivity threshold poses a threat for production with imported raw materials and equipment.
- The lower and upper thresholds of taxation determine the minimum level of taxation required to perform the tasks of the state and the maximum level, which does not hinder the economic development.
- The key macroeconomic task resides in the harmonization of monetary and fiscal policy, which is aimed at economic growth, primarily at the growth of exports and employment. Moreover, the restrictions on normative indicators must be met, and the regulatory indicators must remain within their acceptable bounds.
- The key macroeconomic task makes it possible to overcome the complexity of the systemic representation of the terms ensuring sustainable economic development.
- The art of economic regulation lies in the determination of acceptable intervals for regulatory indicators to overcome the positive sensitivity thresholds of target indicators, given the normative indicators remain in their favorable intervals.

Chapter 5

KEY PRINCIPLES OF ECONOMIC REGULATION

§1. Economic Diseases Caused by Regulation

Consequences of fragile stability

Achieving sustainable price stability is problematic without considering the causal links of prices with the entire financial system. The need to consider those links becomes increasingly evident after each financial and economic crisis.[1] Low inflation, which is artificially achieved by increasing the bank rate, will be accompanied by high reserve requirement, restrained exchange rate, depletion of foreign exchange reserves and high interest rates for government bonds. In those conditions, the seemingly stable annual inflation is followed by unreasonably high real bank rate, seasonal and structural substantial price fluctuations, the decline in GDP growth rate, deterioration of its structure and a decrease in the growth of exports. Even with seemingly stable low inflation, those characteristics certainly are not reminiscent of financial stability, making the very process of ensuring price stability ineffective.

Under those conditions the penalty for artificially low inflation is high, and the risk of its sudden leap increases even from minor changes in the world economy. Paradoxically, low inflation, in that case, becomes very fragile. Extra-systemic price stability arises. Hence, the contradiction between price stability and financial stability may result in sudden price fluctuations, especially for the most sensitive prices of essential goods.

In transforming economies inflation below 2 percent or above 5 percent is usually accompanied by a significantly high real bank rate, which is more than 2 percent. Moreover, a high real bank rate may substantially push up

---

1 T. Adrian and H. S. Shin, "Financial Intermediaries, Financial Stability, and Monetary Policy," *FRB of New York Stuff Report*, 2008; R. McKinnon and G. Schnabl, "The East Asian Dollar Standard, Fear of Floating, and Original Sin," *Review of Development Economies* 8, no. 3 (2004): 331–360.

the risk of uncontrollable inflationary processes. Those bounds represent estimated sensitivity thresholds for inflation.

Exiting the marked interval about the interconnected key economic indicators will stipulate the occurrence of the following economic diseases:

1. A decline in economic growth, deterioration of its structure;
2. Decrease in exports;
3. Reduction of foreign exchange reserves;
4. Growth of public debt, mostly external;
5. Inadequately low monetization ratio;
6. Seemingly stable exchange rate, which is not conducive to GDP growth or improvement of its structure.

The smaller the attainable interval for inflation, the more it contributes to economic predictability and development. Thus, it is better to have sustainable and predictable inflation between 2 percent and 5 percent, that is, an interval of no more than 3 percentage points, than to aim for inflation below 2 percent with a probability of a sudden leap in this indicator. The probability of this event may be particularly high in transforming economies. In this case, excessively low inflation is typically accompanied by a high real bank rate, with the abovementioned concomitant diseases raising the probability of a sudden leap in inflation.

With imports massively exceeding exports, inflation is largely affected by the prices of imported goods. In this case, the pursuit of low inflation, combined with an increase in prices for imported goods, may not only cause an increase in the real bank rate and reserve requirement but also de facto keep the national currency at an artificially overpriced level, which is de jure considered floating. That has to be compensated by cutting down on foreign exchange reserves and monetization ratio having their upper bound of economic security or sensitivity threshold.

Prices often change in steps, rather than gradually. This further increases the risk of deviations from the predicted results. The effect of the structure and seasonality of inflation on its sensitivity thresholds should also be considered. Inflation should not have notable seasonal fluctuations within its interval and still be considered stable even if the annual inflation appears to be such. It is expedient to estimate the sensitivity thresholds for key monetary indicators, which have an extremely high impact on economic agents.

It is better to have reasonable inflation within the acceptable interval with a growing economy than a seemingly stable low inflation and exchange rate with a stagnant economy and a high probability of crisis phenomena that will lead to significant economic problems.

Even if inflation is sensibly 3–4 percent for a transforming economy and 1–2 percent for a developed country, the exceptionally stable exchange rate may negatively affect the balance of payments and the balance of trade. In this case, the goods of developed countries will be more competitive compared to those of the transforming economies, since inflation is lower in developed countries. The rise in prices of imported goods from developed countries will, on average, be 2 percentage points less than the rise in prices in a transforming economy. The export price of goods to the developed countries will also increase by the same amount. This negatively affects exports, especially the exports of finished goods with high added value. To remain within the same competitive terms, gradual national currency depreciation of about 2 percent is advantageous for a transforming economy. Otherwise, the national currency may become overvalued, which creates the basis for a sudden depreciation amid a crisis.

The continuous depletion of foreign exchange reserves during a quarter signals an overvaluation of the national currency. If foreign exchange reserves exit their sensitivity threshold, becoming less than 25 percent of imports, then this implies that the exchange rate has dropped out of the system of monetary indicators. This means that the following unfavorable chain has been triggered in the economy: an overvalued national currency → decrease in exports with a possible increase in imports → sharp deterioration of the balance of trade and the balance of payments → decrease in foreign exchange reserves. This chain of events is not conducive to economic growth and will not ensure the fulfillment of determining budget revenue targets.

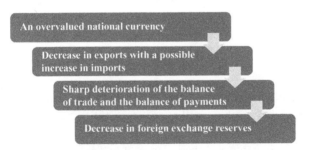

The effect on the economy when the exchange rate is overvalued

Regulatory and normative indicators should have the highest likelihood to remain in their minimal interval because the event of exiting the interval immediately changes the game rules in the economy. For example, if the exchange rate exits its interval, then it will change the conditions for exports, imports and investments. Thus, the outlined method of the minimal interval for key indicators is essential. This method reduces uncertainty in the economy and increases its predictability.

Paradoxically, the rigid stability of the exchange rate increases the risk of its sudden leap. The exchange rate must be considered in a single system with other economic indicators. A flexible exchange rate that is directly linked to the balance of payments and the balance of trade is more predictable than a rigid one with a low probability of long-term stability. Rigidity often crumbles under pressure. The aim to tightly regulate the national currency decreases the scope of its fluctuations; yet, rare fluctuations are often significant, even destructive.

Sudden fluctuations of the exchange rate are caused not by gradual national currency depreciation but rather by the compressed spring effect when the spring bounces after the artificial rigidity that holds it crumbles.

The determination of the acceptable interval for the exchange rate faces certain problems. The exchange rate is formed in the market. However, it is influenced by the bank rate and has a causal link with inflation, the balance of payments and the balance of trade. Needless to say, one should not artificially adhere to a certain regime of the national currency. This may create significant problems for the economy. It is also very problematic if the exchange rate exits its acceptable interval.

Possible issues with the national currency dynamics

Issue 1

A stable national currency has a positive effect on investments, provided interest rates are reasonable. Still, there should be no barriers to the establishment of the true exchange rate for investments. That said, there is an issue of preventing a sudden leap of the exchange rate, which could cause the exit of the currency from its acceptable interval.

Issue 2

The reduction of the bank rate and reserve requirement may result in national currency depreciation. This will stimulate exports, and if imports significantly exceed exports, then this can help improve the situation with both the balance of payments and the balance of trade. However, if the economy is dominated by imports, the currency depreciation could lead to a sharp rise in the prices of imported goods, which may contribute to a significant increase in inflation and the exit from its favorable interval. It is advisable to account for this limitation, because if the economy is significantly dependent on imports, then the exit of the national currency from its permissible interval will spur inflation.

Issue 3

In the event of national currency appreciation, its exit from the sensitivity threshold for exports could affect strongly both the external and internal competitiveness of domestic producers and, as a result, the economic growth.

The national currency exiting its permissible interval with excessive depreciation or appreciation will thereby cause negative effects. Albeit the exporters are understandably interested in gradual depreciation, which will increase the competitive advantage of their manufactured goods when exporting, since a significant part of production costs are in national currency, with the proceeds in convertible currency. On the other hand, appreciation is advantageous for importers and modernization. Thus, the subjective factor of exporters along with the balance of payments and the balance of trade, as well as the entire system of key monetary indicators, will modify the lower interval bound of the exchange rate. Likewise, the subjective factor of importers will modify the upper bound of the exchange rate.

The very floating state of the exchange rate within the permissible interval determines the optimal conditions for the national currency, and it is, of course, an artificial assumption to consider a precise determination of the exchange rate as optimal.

The volatility of economic indicators within their acceptable interval is highly important. Rigid stability is fraught with steep and significant changes in the economy. This is a necessary condition not only for the exchange rate but also for all normative and regulatory indicators within the system of key indicators.

The dangerous collapse of the national currency will occur when the stability of the exchange rate is artificially maintained using a high bank rate with an accompanying high reserve requirement and, as a consequence, a low monetization ratio. All this is usually followed by a decrease in foreign exchange reserves and an indifference for the condition of the balance of payments and the balance of trade. The artificially accumulated potential for instability is triggered under the pressure. Rigid stability crumbles, and the indicator breaks out of its sensitivity threshold. A permissible and seemingly stable national currency is shown in Figures 6 and 7.

It is only possible to predict the interval conducive for economic development, beyond which, on the one hand, unfavorable conditions emerge for importers with the danger of an inflationary chain, and on the other, highly unfavorable conditions arise for exporters with serious problems for economic development.

The probability of the compressed spring effect for the national currency, leading to higher inflation, will increase sharply if the sensitivity threshold of

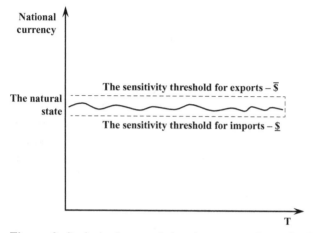

Figure 6 Snake-in-the-tunnel situation, a natural state for the national currency.

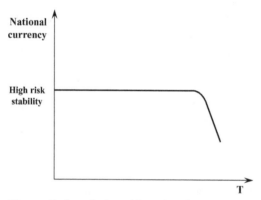

Figure 7 Seemingly stable national currency with a high risk of a sudden leap.

foreign exchange reserves is crossed and they fall below 25 percent of imports, given the overvalued national currency, which is often accompanied by non-systemic rigid price stability. Here the result is the same as with an excessively weak currency. And the harsh outcomes in both cases are identical.

An exchange rate can be truly considered as floating if the foreign exchange reserves are used infrequently as a countermeasure, with no risk of their depletion below 25 percent of imports. At the same time, the proportions of using the foreign exchange reserves are also important. If the level of foreign exchange reserves is sufficient, but the rate of monthly decline exceeds 1 percent for a substantial period, then this also signals that the exchange rate does not correspond to the real state of the economy and has dropped out of the system of key economic indicators.

Gradual currency depreciation does not induce economic disturbance, since it allows to predict the exchange rate within a fairly small window. A potential economic disturbance arises when there is a high probability of the compressed spring rebound with a seemingly stable exchange rate outside the system of key economic indicators. A sudden leap of the national currency is the worst alternative for any economy. In such a situation, the homeostasis of the entire economic system is disrupted, and the likelihood of stable economic growth is practically depleted.

In conditions of uncertainty, the nonsystemic stability of the national currency increases the likelihood of its significant alteration, even with a minor crisis. Former Federal Reserve chair Alan Greenspan noticed: "Uncertainty is not just a pervasive feature of the monetary policy landscape; it is the defining characteristic of that landscape."[2]

It should be noted that the national currency depreciation contributes to the growth of exports and the improvement of the balance of payments and the balance of trade. In turn, the growth of exports, increasing the flow of convertible currency, contributes to the stability of the national currency. This naturally triggers the cyclical effect for the national currency, provided inflation remains in its favorable interval, without exiting its sensitivity threshold.

The closer the national currency is to its lower interval bound, the more it contributes to the growth of exports and import substitution. Overall, this is aimed at improving the state of the balance of payments and the balance of trade.

Currency appreciation contributes to the growth of imports; however, an increase in imports, therefore an increase in the trade deficit, results in the higher likelihood of a sharp depreciation of the national currency, even if it seems to be stable.

The periodical currency appreciation in transforming economies is usually explained by the unjustified Dutch disease. The national currency mainly appreciates not due to the inflow of convertible currency, including remittances, but rather due to the disproportionately tight monetary policy, high bank rate, and, therefore, the low monetization ratio. While securities transactions are not always used to regulate the money supply.

A decrease in the variance of the national currency can only be justified by an improvement of the balance of payments and the balance of trade,

---

2 Alan Greenspan, "Risk and Uncertainty in Monetary Policy," remarks at the Meetings of the American Economic Association, San Diego, January 3, 2004 (and published in the *American Economic Review: Papers and Proceedings* 94 (May 2004): 33–40).

employment and substantial economic growth. With no such prerequisites, the likelihood of a sudden leap for the seemingly stable exchange rate increases.

Given the factor of uncertainty, it is expedient to do a systemic analysis of economic interconnections, including those relating to the exchange rate dynamics. Only the interconnected consideration of the uncertainty factor and economic patterns gives acceptable results.

A systemic assessment of the causal links of the balance of payments and the balance of trade with the exchange rate contributes to the effective regulation of economic processes. For within the unified system of monetary indicators, the exchange rate is directly interconnected with the bank rate, as well as with the reserve requirement, the monetization ratio and, of course, with inflation.

Without considering other key indicators, it is inexpedient to secure the stability of consumer prices and to control asset price inflation, as well as to regulate the national currency. Here the following causal chain takes place. The bank rate corresponding to the inflation in its favorable minimal interval → contributes to the growth of GDP and the improvement of its structure → which in turn enables the exchange rate to be as close as possible to its lower bound → improving the balance of payments and the balance of trade, as well as the employment. This chain is most consistent with the "Magic square," which is a concise form that represents the causal links of key economic indicators. The most brilliant chess combination can be destroyed by rearranging the moves. It is advisable to adhere to this principle in economics as well.

The minimal uncertainty interval can in no way be shrunk into a point. A rigid forecast leading to an attempt to ensure rigid stability is dangerous and fraught with irreparable economic losses. An example can be inflation of less than 2 percent or de facto pegged exchange rate, which does not correspond to the reality of transforming economies. Such a policy often leads to a significant drop in GDP and degradation of the economic structure. With a point forecast, the economy is adjusted to the chosen model and not vice versa. This in effect turns the model into a Procrustean bed for the economy.

§2. Ten Principles for Forecasting Economic Processes

The factors affecting the effectiveness of forecasts
or how to make predictions acceptable

Economic studies and forecasts cannot achieve 100 percent accuracy; however, they can deliver acceptable results. This paragraph systematizes those principles that can significantly increase the efficiency of forecasts and decision making in the face of uncertainty in economic processes.

Principle 1

When making forecasts, it is imperative to consider the data not only from statistical services but also from the stock exchange.

The reasons are as follows:

a) Stock exchange data result from collective decisions.
b) The responsibility for making decisions in the stock market is very high, for wrong decisions may be irreparable. Stock investors risk their own money.
c) Stock exchange data is incomparably more expedite than the data of statistics agencies. It takes time to process the data from those agencies, during which substantial changes may occur.

Stock exchange data is also synthesizing, which primarily reflects the key indicators of the economy. Countries, where the stock exchange is but a formal institution, are deprived of the opportunity to use the exchange data for the real assessment of the economy. Forecasts for such countries are inherently vulnerable.

Principle 2

When planning it is reasonable to reflect forecasts of several experts, among whom there should be not only theoretical but also experimental economists, especially stock investors. It should be borne in mind that forecasts may differ significantly, and even if the discrepancies in the forecasts are insignificant, this does not yet confirm their accuracy.

If the forecasts do not systematize economic interdependencies, the domino effect may set off. The assessment of a single economic indicator, without considering its causal links with other key indicators, may trigger a chain of negative consequences. A bundle of incoherent and hasty measures is inherently ineffective.

Causal links may change over time. The forecasting fitness increases if those changes are considered and estimated, for one cause can have a chain of several effects, just as one effect can have several causes. The principle of interdependence is a key part of the economic system, and causal links here are not rigid or set in stone. If the relation between two indicators is given by a figure, without any obvious causal link, then it is better to recognize such a figure merely as a result of correlation. Correlation does not imply causation. Correlation without causation is usually random.

The forecast is acceptable for the formation of the minimal interval of an indicator if the forecast:

- provides a systemic presentation of causal links for the key economic indicators;
- incorporates the possible dynamic alterations of those links;
- assesses the impact of indicators that are not presented as key.

When estimating the minimal interval for an indicator, the quality of the forecast goes down if those bullets are not met. Such a forecast should be considered less significant or not considered at all. A forecast cannot be applied without the synthetization of the results of experimental data and the identification of the economic patterns to the full extent.

Principle 3

Evaluation of economic processes by means of several quantitative models helps to reduce the minimal uncertainty interval for an indicator and improve the quality of a forecast.

It is advisable to implement those models whose limitations are known, rather than those models whose shortcomings are unknown or difficult to assess. Knowing the strengths and weaknesses of a model is an essential prerequisite for economic research.

Given the differences in the feedback strength of economic indicators, one should consider the conditionality of both the presented causal links and the models used. The knowledge, or rather the recognition of the model limitations, certainly contributes to the formation of a systemic approach to the study of economic processes.

Models are just one of the forecasting tools. It is unproductive to utilize models without the awareness of economic patterns and possible alterations of the intervals of key indicators. All models are conditional in one way or another, and the formation of uncertainty intervals makes it possible to reduce the level of their conditionality.

Principle 4

It is important to present the minimal intervals of the uncertainty of key indicators for different probabilities. Furthermore, the study should be systemic. It should consider the possible impact of the change of one minimal interval on the interval of another key indicator. Interval determination is a similar job to sweeping mines, for the area outside the interval, though having low probability, may have irreversible consequences.

The determination of the minimal interval for an economic indicator should include safety measures in case of significant economic changes, as well as a mechanism for its adjustment. The assessment of the interval for an indicator should be done in conjunction with safety measures, in case of exiting said interval. This helps to implement effective economic decisions.

Principle 5

It is useful to define at least two intervals of key economic indicators. The first, smaller interval should include the most probable values at which the economic system remains stable or at least can neutralize the possible threats. In doing so, the probability assessment of sensitivity thresholds for the intervals of the key target, normative and regulatory indicators, has a decisive role.

The second, bigger uncertainty interval should include the unlikely events or the known unknowns. Those describe economic scenarios with an assessment of anticipated hypothetical harsh situations, whose impact on the economy is difficult to determine, though it may be quite substantial. As for the extremely unlikely events that cannot be anticipated and are outside the second uncertainty interval, it is almost impossible to predict anything there. It is impossible to predict the unknown unknowns. Although the development of certain scenarios can contribute to the development of airbags for extreme situations, that cannot be anticipated.

The following approach should be implemented:

- Ninety percent probability with a narrower minimal interval, which is more appealing. Though, it should be borne in mind that values outside the interval can have a significant impact.
- Ninety-five percent probability with a wider interval. In this case, the probability of exiting the interval is reduced.

It is unreasonable to strictly adhere to the forecast, and this applies both to 95 percent and even more to 90 percent probability. The forecast results should be implemented with caution, for the economy is volatile and the impact of exiting the interval can be difficult to estimate. Any forecast must be complemented with a set of preventive and safety measures.

Principle 6

Extremes should be avoided when determining the minimal uncertainty interval. A reduction in the uncertainty interval is acceptable only without significantly decreasing the likelihood of its occurrence. A point forecast is initially not feasible. On the other hand, an extremely wide interval with almost 100 percent probability is tautological and is meaningless for the forecast.

For a key indicator, the interval can become meaningless if it approaches a point estimate because a small probability of occurrence is of little value. At the other extreme, the interval provides no meaningful information if it is broad with an unreasonably high probability. This interval is of a little value as well. As often happens, the extremes are close in meaning and produce no results.

In general, the interval of an indicator corresponds to its probability. The higher the probability, the larger the interval, and, vice versa, the smaller the interval, the lower the probability. When identifying the minimal interval one should, whenever possible, distinguish the conditions when the value of the key indicator will exit its sensitivity threshold and what may be the cause of it. Interval diagnosis is no less valuable than an assessment of the likelihood of its implementation.

Principle 7

If the expected consequences are insignificant when exiting the interval bound, then it is reasonable to narrow the minimal uncertainty interval. In this case, the possible impact or the probability multiplied by the force of its impact does not change much. Still, the consequences are assessed considering both the probability of the event and the force of its impact caused by its occurrence. The more the possible impact, the more careful one should be in forecasting.

Hence, if there are insignificant consequences when exiting the interval bounds of an indicator, it is advisable to reduce the minimal uncertainty interval for the said key indicator given the increase of possible impact outside the interval is negligible and the quality of the interval does not decrease. It should be noted that the assessment of insignificance for various indicators is different.

Principle 8

The minimal uncertainty interval does not need to be widened if the probability of an event outside the interval is extremely small, though its consequences may be disproportional. Nobody spends huge funds on specially equipped bomb shelters in the case of a nuclear war. In any case, the likelihood of the total hell is negligible. However, it should be noted that in different conditions the possible impact may alter for unaccounted negligible probabilities, and the probability itself may also change. Thus, herein the economic system requires a dynamic analysis. It should be noted that the determination of the dynamics for uncertainty intervals under different scenarios, especially the most probable ones, increases the quality of forecasts. Simulation is helpful here.

Principle 9

The long-term economic forecast is quite conditional and can be formed only in a rather wide interval. As a rule, short- and medium-term interval forecasts, which require regular adjustments, are of practical importance.

In practice, forecasts should have to expand the uncertainty band and require periodic adjustments (Figure 8).

The wider uncertainty band has a forecast probability of 95 percent, while the narrower band has 90 percent. It should be borne in mind that over time the minimal interval of the band may expand so much that after a certain period it will lose its value. It is unlikely to obtain a reasonable long-term forecast.

Principle 10

It is necessary to correct forecasts at least every six months, especially with increased volatility, since the uncertainty intervals are in expanding bands. Note that adequately forecasting each subsequent period is more difficult than the previous ones combined.

The study of how well the minimal uncertainty intervals function in the real economy, rather than only in theoretical models, helps to improve their quality.

If any one of those principles is not met, then the forecast becomes more conditional and the probability of forecast realization, as well as its quality, is significantly reduced. It is quite possible to considerably reduce the minimal

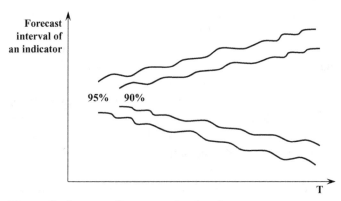

Figure 8 An expanding uncertainty band.

forecast interval with the help of the presented principles. The study should also indicate which predictions are partially feasible, what kind of reservations must be met in that case and what consequences can be considered and assessed.

The economy cannot be precisely predicted, but it is certainly possible to be better prepared for the unexpected if its likelihood and consequences are considered. Economic regulation is also the art of the possible.

Summary of Chapter 5

§1. Economic diseases caused by regulation

- Exiting the interval of economic indicators, favorable for economic development, will stipulate the occurrence of the following economic diseases: decline in economic growth, deterioration of its structure; decrease in exports; reduction of foreign exchange reserves; growth of public debt; inadequately low monetization ratio; and inconstant exchange rate.
- Low inflation, which is artificially achieved by increasing the bank rate, will be accompanied by high reserve requirement, restrained exchange rate, depletion of foreign exchange reserves and high-interest rates for government bonds. In those conditions, the seemingly stable annual inflation is followed by unreasonably high real bank rate, seasonal and structural substantial price fluctuations, the decline in GDP growth rate, deterioration of its structure and a decrease in the growth of exports. Low inflation in that case becomes very fragile.
- If inflation is largely affected by the prices of imported goods, then the pursuit of low inflation, combined with an increase in prices for imported goods, may not only cause an increase in the real bank rate and reserve requirement but also de facto keep the national currency at an artificially overpriced level, which is de jure considered floating. This increases the risk of deviations from the predicted results.
- The rigid stability of the exchange rate increases the risk of its sudden leap. The exchange rate must be considered in a single system with other economic indicators. A flexible exchange rate that is directly linked to the balance of payments and the balance of trade is more predictable than a rigid one with a low probability of long-term stability. Rigidity often crumbles under pressure. The aim to tightly regulate the national currency decreases the scope of its fluctuations; yet, rare fluctuations are often destructive.
- The dangerous collapse of the national currency will occur when the stability of the exchange rate is artificially maintained using a high bank rate

with an accompanying high reserve requirement and, as a consequence, a low monetization ratio. All this is usually followed by a decrease in foreign exchange reserves and an indifference for the condition of the balance of payments and the balance of trade. The artificially accumulated potential of instability is triggered under the pressure.

• A snake-in-the-tunnel situation is presented, which describes, on the one hand, unfavorable conditions for importers with the danger of an inflationary chain, and on the other, highly unfavorable conditions for exporters with serious problems for economic development.

• A rigid forecast leading to an attempt to ensure rigid stability is dangerous and fraught with irreparable economic losses. With a point forecast, the economy is adjusted to the chosen model and not vice versa. This in effect turns the model into a Procrustean bed for the economy, which often leads to a significant drop in GDP and degradation of the economic structure.

§2. Ten principles for forecasting economic processes

• When making forecasts, it is imperative to consider the data not only from statistical agencies but also from the stock exchange.

• When planning it is reasonable to reflect forecasts of several experts, among whom there should be not only theoretical but also experimental economists, especially stock investors.

• Evaluation of economic processes by the means of several quantitative models helps to reduce the minimal uncertainty interval for an indicator and improve the quality of a forecast.

• It is important to present the minimal intervals of the uncertainty of key indicators for different likelihoods.

• It is useful to define at least two intervals of key economic indicators.

• Extremes should be avoided when determining the minimal uncertainty interval.

• If the expected consequences are insignificant when exiting the interval bound, then it is reasonable to narrow the minimal uncertainty interval.

• The minimal uncertainty interval does not need to be widened if the probability of an event outside the interval is extremely small, though its consequences may be disproportional.

• The long-term economic forecast is quite conditional and can be formed only in a rather wide interval.

• It is necessary to correct forecasts at least every six months, especially with increased volatility, as the uncertainty intervals are in expanding bands.

CONCLUSION

The interval determination of indicators is a method of research for precisely indeterminate indicators under the state of uncertainty natural for the economy. The interval representation of the indicator enables a description to be made of its most likely values. The interval method of economic indicators will open a logical path from the predicament of explaining its quantitatively precise indefinability and deploying a complete picture of causal links in the economy.

The relations of economic indicators can be represented using mathematical models. However, the most detailed record of economic links will not change the fundamental impossibility of determining the exact values of economic indicators. Considering the principle of uncertainty, the causal links of key indicators can be presented by defining their range of possible values. The interval uncertainty of key economic indicators represents the infeasibility to pinpoint an economic indicator within the minimal uncertainty interval.

The difference between the lower and upper bounds of the key indicator is its minimal uncertainty interval for a given probability. This range is the T-interval.

When forecasting key economic indicators, the use of the minimal uncertainty interval is an underlying idea. Based on this idea, a system of intervals of key economic indicators can be deduced, which is a necessary condition for economic efficiency. Thus, economic predictability can be maximized.

Scenario forecasts can reduce the uncertainty interval of an economic indicator in each scenario. However, they cannot eliminate uncertainty by turning the interval into a point estimate. An overlap can be purely accidental, like a shot in the dark. Decision making which is subject to uncertainty amounts to the determination of the most possible minimal uncertainty interval for key indicators and designing scenarios for economic development with their implementation methods.

The diagnosis of the minimal interval of an economic indicator, including its extreme values, and possible consequences when obtaining such values, is no less important than the forecast itself. When carrying out an interval forecast of an indicator, its diagnosis is also necessary. The utility of the forecast depends on the interval bounds—the less, the better. It also depends on how accurate the estimates of the possible consequences in case of obtaining results outside the interval are. Depending on the interval bounds of the key indicator, the corresponding airbags should be designed.

Sensitivity thresholds essentially are critical bounds beyond which the likelihood of unforeseen negative and substantial events increases significantly. Moreover, those bounds are dynamic and can change over time. To assess economic homeostasis, it is advisable to evaluate the thresholds that lead to a negative impact of quantity on the quality of economic growth. The sensitivity threshold of an economic indicator outlines the possible consequences when exiting the bounds of the uncertainty interval. Beyond the sensitivity threshold, the elasticity of an economic indicator can undergo a phase transition, as at this critical stage the indicator becomes very sensitive.

The volatility of economic indicators within their acceptable interval is highly important. Rigid stability is fraught with steep and significant changes in the economy. The artificially accumulated potential of instability is triggered under the pressure.

The necessary conditions for an efficient economic system are the determination of possible intervals of a key target, normative and regulatory indicators; the systematization of key indicators, their presentation in the form of a single macroeconomic task; the identification of possible methods of regulating the economic indicators, including the inefficient ones; the elaboration of rational methods of regulating economy under uncertainty; and the adoption of efficient solutions, matching the conditions of economic performance, as well as their evaluation.

Regulatory and normative indicators should have the highest likelihood to remain in their minimal interval since the event of exiting the interval immediately changes the game rules in the economy.

The presented intervals of a target, normative and regulatory indicators; the formulated key macroeconomic task; and the 10 principles of forecasting contribute to a real assessment of the economy in its natural conditions of uncertainty. Fulfilling the constraints of key economic indicators, in a unified system, is a prerequisite for the development of the economy.

Economic regulation is the art of the possible. The art of economic regulation is the determination of acceptable intervals for regulatory indicators. This is necessary to optimize the target indicators when the condition of finding normative indicators in a favorable interval is fulfilled. The adjustment of regulatory indicators in their acceptable intervals promotes the development of the economy and stabilizes the situation.

The difference between the upper and the lower bounds of regulatory indicators determines the scope of the art of regulation influencing the economic processes. The art of regulation is defined as finding an effective solution within this scope.

The interval method for analyzing economic indicators contributes to a systemic presentation and a realistic assessment of GDP growth and the possible range of changes in exports, imports and employment. This method helps to identify the favorable intervals of inflation, budget deficits, public debt, monetization ratio and foreign exchange reserves, as well as assess the permissible intervals of nominal and real bank rates, reserve requirement, national currency and rates of taxation.

Uncertainty intervals are a must for economic forecasting and regulation. The systemic presentation of intervals of key indicators ensures the implementation of effective decisions.

APPENDIX

The Uncertainty Relations of Economic Indicators

The precise definition of economic indicators is impossible in any model, even if the model represents the economy in an extremely simplified form. It is only possible to formulate their ratio in general terms.

Illustrative examples of models are used herein since the complexity of the model is not critical for describing generalized functional interdependencies between indicators.

Below is an example of a simplified quantitative economic model:

$$\sum_{j=1}^{n} g_j q_j \rightarrow \max$$

$$\sum_{j=1}^{n} a_{ij} q_j \leq R_i \quad i = \overline{1,m}$$

$$q_j \geq 0 \quad j = \overline{1,n},$$

where g_j is the share of income in the price of the product j, a_{ij} is the expenses of the i-th resource in the unit production of the j-th product, R_i is the volume of the i-th resource and q_j is the volume of product j.

The dual problem of the considered model is formulated below:

$$\sum_{i=1}^{m} R_i u_i \rightarrow \min$$

$$\sum_{i=1}^{m} a_{ij} u_i \geq g_j \quad j = \overline{1,n}$$

$$u_i \geq 0 \quad i = \overline{1,m},$$

where u_i is the optimal valuation of the i-th resource.

From the duality principle the optimal values satisfy the following conditions:

$$\left[R_i - \sum_{j=1}^{n} a_{ij} q_j\right] \cdot u_i = 0 \qquad i = \overline{1, m}$$

$$\left[g_j - \sum_{i=1}^{m} a_{ij} u_i\right] \cdot q_j = 0 \qquad j = \overline{1, n}.$$

In the primal problem, the volumes of resources R_i, the share of income g_j, the objective function of the direct problem and technological rates of resource consumption are known. Resources u_i and volume of output q_j are unknown and to be determined. It can be observed from the obtained ratio that if u_i and q_j are known, then the values of R_i and g_j can be determined. Therefore, for any given set of variables, the optimal values of others can be obtained.

Thus, if a model contains price variables g_j and u_i while it is necessary to determine quantitative variables q_j and R_i, then separately for both quantitative and price variables $m + n$ linear equations with $m + n$ unknowns can be obtained. Provided that the coefficients of the unknowns are positive, the rank of the matrix composed of the coefficients of the unknowns will be equal to $m + n$; in so doing, the quantitative variables will be found. Under the same condition, if quantitative variables are given, then price variables will be found.

Likewise, in any optimization model, point estimates of one group of quantitative and price variables can be found if only the other group is assigned a value. To determine some variables, the values of other variables must be taken for granted in advance.

The relations of presented quantitative and price indicators do not infer that the change in the volume of output is necessarily the cause of change in price indicators since their causality is not always unambiguous.

Similar ratios can be obtained when choosing the main target indicator for GDP growth, exports or employment. Similar ratios can be obtained for nonlinear models.

Below is a generalized optimization model of economic growth:

$$\text{GDP}(Q) \to \max$$

$$f_i(Q) \leq R_i \qquad i = \overline{1, m}$$

$$Q \geq 0$$

The model requires resource constraints to be met.

The Lagrange function for the general optimization model for economic growth is as follows:[1]

$$L(Q,U) = GDP(Q) + \sum_{i=1}^{m} u_i \left[R_i - f_i(Q) \right].$$

The relations below follow the Kuhn–Tucker conditions for optimal economic indicators:

$$[R_i - f_i(Q)] u_i = 0 \qquad i = \overline{1,m}$$

$$\left[\frac{\partial GDP(Q)}{\partial q_j} - \sum_{i=1}^{m} \frac{\partial f_i(Q)}{\partial q_j} u_i \right] q_j = 0 \qquad j = \overline{1,n}.$$

Given that $m + n$ quantitative variables of GDP and resource constraints are known, it would seem possible to determine their price variables and vice versa. However, in optimization models that characterize generalized causal links of economy, quantitative variables only formally define price variables. The converse holds as well. Moreover, this is true for any combination of variables.

---

1 A. A. Tavadyan, Uncertainty Intervals of Economics (Moscow: Nauka, 2012).

ACKNOWLEDGMENTS

First of all, let me express my gratitude to my colleagues. The experience of working with them both in the political and economic fields made it possible to formulate the main principles of this book. Not only my investigations on the whole variety of economic processes but also the advice from many of my friends and colleagues gave me some food for thought.

The conception of the book is facilitated by the attention and care from the people who are close to me. I am thankful to my family, who created all the conditions for me to work on this book. I want to thank my wife Sona, who was attentive to me during my long working sessions.

I want to also thank all the people who have read the manuscript. In the final stages of this journey, I had valuable help from the staff members at Anthem Press. I am thankful to them for their guidance.

Separately, let me express my gratitude to Aghasi Tavadyan for the advice and help in writing the book. He has been indispensable with his erudition and suggestions, which dot the book. He also gave the cover idea for this book. I am thankful to Vladimir Khachaturyan, who provided helpful edits.

INDEX

acceptable intervals 5, 47, 50, 57–60, 62, 80
accumulated problems 4, 17, 24
Achuthan, Lakshman 41
Act to Promote Economic Stability and Growth 42
Adrian, Tobias 63
American Economic Association 69
annual government deficit 43
Armstrong, Scott 12
art of economic regulation 5, 59, 62, 80
Austria 46

balance of payments 19, 46–47, 51–53, 56, 65–67, 69–70, 76–77
balance of trade 38, 42, 46–48, 52 53, 56, 58–59, 61, 65–67, 69–70, 76–77
Banerji, Anirvan 41
bank rate 19, 38, 42, 49–55, 58–59, 62–64, 66–67, 69–70, 76
Basu, Kaushik 55
business cycles 3, 34, 39

Canada 51
causal links 2–4, 6, 8–10, 12–14, 19–20, 22–23, 25, 35–36, 48, 63, 70–72, 79, 85
causality 1–2, 4, 6, 10, 22, 84
causation 8–11, 13–14, 41–42, 60–61, 71
cause-and-effect relations 17–19
China 45–46
competitiveness 55, 57, 67
complexity of interconnections 13
compressed spring effect 14, 17, 24, 66–67, 69
confidence interval 27–29
constructive thinking 14
correlation 18, 31, 71
crises 3, 13, 19, 32, 34, 38–39, 42, 61
critical bounds 35–36, 80
critical situations 10, 31, 37–38, 50

currency 32, 36–38, 49–51, 53–56, 62, 64–70, 76, 81; appreciation 55–56, 67, 69; depreciation 36–37, 53–56, 62, 65–67, 69

Dadayan, Vladislav 21
Dawes, Robyn 12
Denmark 46
developed countries 42, 44–45, 47–49, 51–52, 54, 58, 62, 65
domino effect 71
Dutch disease 69
dynamic nature of economy 2, 8, 13, 18–19, 28, 36, 74, 80

economic airbags 29, 31, 33, 38, 73, 80
economic assessment 4–5, 13, 15, 18–20, 24, 28, 32, 36–37, 40, 59, 70, 71, 73–74, 80–81
economic benchmarks 38, 40, 45, 54
economic bottlenecks 13, 15
economic causalities 1, 4
economic connections 2–3, 6, 8–11, 13–15, 23, 44, 60
economic cornerstones 3, 11–12
economic cycles 34
economic diagnosis 5, 18, 32, 74, 79
economic diseases 5, 63–64, 76
economic game 17, 24, 59, 65, 80
economic growth 1, 4, 13, 35–36, 40–46, 48–62, 64–65, 67, 69–70, 76, 80, 84–85
economic homeostasis 4, 36, 40, 69, 80
economic indicator 3, 5, 7, 10, 17–19, 23, 28–29, 31–32, 34–37, 39–40, 59–61, 64, 66–73, 76–77, 79–80; distribution of economic indicator 29; uncertainty of economic indicators 2, 27, 72, 77, 79
economic interconnections 10, 13, 23, 25, 70

economic interdependencies 1, 3, 8–9, 11, 15, 22, 71, 83
economic landscape 2, 36
economic mindset 13–15
economic modeling 2, 8
economic objective 5, 9, 42, 51, 53, 60–61
economic patterns 10, 12, 18, 20, 24, 33, 38, 70, 72
economic policy 3–4, 32, 35, 38, 41–42, 47, 52, 61
economic potential 38, 40, 49–50
economic predictability 64, 66, 79
economic processes 1–2, 4–15, 18–25, 27, 33–34, 38–39, 41, 50, 59–60, 70, 72, 77, 79, 81
economic regulation 2, 8–10, 41, 45, 50, 57, 59–63, 65, 67, 69, 71, 73, 75–77, 81
economic research 1–2, 6, 8–9, 11–13, 15, 18, 41, 72, 79
economic security 36, 38, 40, 64
economic stability 3, 38, 40–43, 46, 48, 76
economic system 1, 3–4, 8–15, 22–23, 25, 32–33, 35, 39, 41–43, 48–61, 65–71, 73–74, 76, 79–80; single economic system 10, 14, 48, 61
economic uncertainty 59
effect of extrapolation 18
Einstein, Albert 7
emergency measures 31, 38
employment 19, 35, 38, 41–43, 46–47, 58–59, 61–62, 70, 81, 84
Eurasian Economic Union 43
European Central Bank 43
European Monetary System 44
European Union 43–44, 51
exchange rate 20, 43–44, 47–48, 52–56, 58, 62, 64–70, 76
expensive money 54
exports 35, 37–38, 41–43, 46–47, 51–59, 61–69, 76, 81, 84

factor of uncertainty 1, 9–10, 12–13, 17–20, 24, 35, 70
favorable intervals 5, 59, 61–62, 81
fiscal indicators 50, 57, 61
fiscal policy 48, 58, 61–62
forecast efficiency 5, 30, 38
forecast horizon 31
forecast quality 3, 34, 39, 59
forecasting methods 17, 31

foreign exchange reserves 38, 42–43, 48–49, 51–53, 58, 63–65, 67–68, 76–77, 81
foreign trade 42–43
formalization 10
fragile stability 32, 63, 76

game rules 59, 65, 80
Gaussian curve 28
GDP 5, 19, 29, 35–38, 41–42, 44–47, 49, 51, 57–59, 61–64, 70, 76–77, 81, 84–85
German reunification 46
Germany 41
Gigerenzer, Gerd 12
Goldman Sachs 20
Goldstein, Daniel 12
Goodhart, Charles 42
government debt 38, 43–44, 49, 58, 64, 76, 81
government deficit 43–44, 49, 58
Greenspan, Alan 69

Hatzius, Jan 20
Hibon, Michele 12

illusion of optimality 21
illusion of stability 3, 27
implemented policies 17
imports 46, 48–49, 55–56, 58, 62, 64–66, 68–69, 81
indicator bounds 1, 14, 30–33, 35–38, 40, 43, 47–48, 50, 56, 58, 60, 62, 64, 74, 79–80
inflation 19, 28, 32, 36–38, 42–48, 50–56, 58, 62–67, 69–70, 76, 81; cost-push inflation 53; inflationary chain 67, 77
insurance measures 23
interest rate 44, 49, 53
interval bounds 32–33, 50, 56, 74, 80
interval determination 8, 72, 79
interval forecast 3–4, 13–15, 19, 24, 28, 75, 79
interval method 2, 7–9, 13–14, 79, 81
interval representation 8–9, 79
International Monetary Fund 49
investments 37, 51, 53, 65–66
Ireland 46

key causal links 9
key indicators 3, 5, 9, 11, 14, 22–23, 25, 30–33, 35, 37–39, 41–43, 48, 50, 57, 59–60, 66–67, 70–72, 77, 79–81; bounds of key indicators 31, 38

key macroeconomic task 5–6, 14, 38, 50, 58, 61–62, 80
Keynes, John Maynard 2
Kuhn-Tucker condition 85

Lagrange function 85
logical structuring 10, 14
long-term programs 33
low-validity system 1, 33–34, 39
Loungani, Prakash 29

Maastricht criteria 43–44, 49
Maastricht Treaty 43, 49
Magic square 41–42, 70
Makridakis, Spyros 12
Master, Loretta 21–22
mathematical models 19, 24, 79
McKinnon, Ronald 63
McNees, Stephen 21
minimal losses 7
minimal uncertainty intervals of economic indicators 3, 13, 15, 27, 29, 31–33, 35, 37–39, 59–60, 70, 72–74, 77, 79
model assumptions 11
monetary indicators 50, 57, 61, 64–65, 67, 70
monetary policy 32, 41–42, 48, 50–53, 61–63, 69
monetization level 38
monetization ratio 42–43, 46–55, 58, 62, 64, 67, 69–70, 76–77, 81
money supply 37, 69
most probable distribution of an economic indicator 29

national currency 32, 36–38, 49–50, 54–56, 62, 64–70, 76, 81
normal distribution 28
normative indicators 5, 43, 47, 58–60, 62, 65, 80

Occam's razor 7
optimization models 23, 85

past-based forecasts 17
permissible threshold 57
phase transition 4, 35–36, 40, 80
point estimate 7, 19, 27, 29, 31, 74, 79
point forecast 5, 70, 73, 77
policy coordination 48, 61
possible scenarios of an uncertainty band 35

precise forecast 3, 19, 24, 32, 38
precise measuring 7
precisely indeterminate indicators 8, 79
price stability 41–44, 48, 61, 63, 68
primary economic objective 42, 61
principle of uncertainty 14, 79
principles of systematization 14
probability 18–19, 24–25, 27–32, 35, 37–39, 42, 46, 52, 55, 59, 61, 64, 66–67, 69, 72–77, 79
Procrustean bed 5, 11–12, 21, 24, 70, 77

quackery 30
qualitative methods 8
quality of research 8
quantitative analysis 6, 8, 19
quantitative changes 8, 11, 12
quantitative methods 11, 72, 77
quantitative precise indefinability 4, 7, 14, 79
quantitative thinking 7, 11, 15

rapidly developing countries 45, 51
real bank rate 50–55, 58, 62–64, 76
regulatory indicators 5, 18, 44, 47, 50, 58–62, 67, 73, 80
research method 13, 15
reserve requirement 38, 42, 50–54, 58, 63–64, 66–67, 70, 76 77, 81
rigid forecast 5, 70, 77
rigid stability 5, 17, 24, 66–67, 70, 76–77, 80
risks 13, 15, 28

scenario analysis 3, 19, 24
scenario forecast 31, 79
Schnabl, Gunther 63
sensitivity thresholds 2, 4–5, 14, 18, 29, 35–40, 44–45, 47–49, 51–59, 61–62, 64, 67–69, 73, 80; negative sensitivity thresholds 35; positive sensitivity thresholds 35
Shin, Hyun Song 63
Silver, Nate 20–21, 30
Singapore 45–46
snake-in-the-tunnel situation 68, 77
South Korea 45
spring effect 66–67
stagnation 45
state of equilibrium 1
statistical services 19, 71
Stiglitz, Joseph 52
stock exchange 19, 71, 77

Switzerland 46
system causations 11
system of key economic indicators 1, 3,
    8–11, 14–15, 22, 25, 33, 60, 65–67,
    70, 79
system structure 11
systematization 1–2, 6, 8, 10, 14, 35, 39,
    41, 60, 80
systemic analysis 1, 10, 13, 70
systemic approach 5, 9–10, 14, 42, 48,
    60, 72
systemic thinking 11
systemology 3, 7–8, 10, 14

Taleb, Nassim 12, 20
target indicators 5, 18, 41, 43–44, 47, 50,
    58–62, 80
tautology 30
taxation 38, 56–58, 62, 81
ten principles of forecasting 6, 33, 80
T-interval 31, 38, 79
trade deficit 36, 69

transforming economies 43, 45, 47–54,
    58, 62–65, 69–70
Treaty on the Eurasian Economic Union
    43–44, 49

uncertainty bands 1–4, 6–8, 10, 12, 14,
    17–18, 20, 22, 24, 27–28, 30, 32–36,
    38–42, 62–64, 66, 68, 70, 72, 74–76,
    79–80, 83–84; expanding uncertainty
    band 4, 75
uncertainty intervals 3–5, 12–13, 15, 18,
    24–25, 27, 29–30, 33, 38–39, 41, 72,
    74–75, 77–81, 85
uncertainty relations 12, 14, 21–25, 83
unemployment 36–37, 48
unforeseen events 1, 19, 24
unpredictability 34
USA 41, 51

variability of the situation 31
Varoudakis, Aristomene 55
Voltaire 22

CPSIA information can be obtained
at www.ICGtesting.com
Printed in the USA
JSHW050149020622
26614JS00001B/6